— 150 — VINEYARDS YOU NEED TO VISIT BEFORE — YOU DIE —

By Shana Clarke

Lannoo

Introduction By Shana Clarke

When I first started this book, I had intended for it to be an index of the world's most iconic vineyards that would highlight plots of land which produced beautiful, jaw-dropping wines. Famous names. The grandest estates. But the more I thought about the project, the less this approach made sense. The wine world is so much more than just well-known names and sites. Wine is produced in all kinds of surprising places, and the joy of discovering them can be akin to drinking a great glass of Burgundy, Barolo or port.

So, I shifted gears. I began thinking about wine in the bigger global context. Given the large number of grape-growing areas in the world, why not shine a spotlight on places that aim to cultivate their own Grand Cru vineyards?

I realized there is no standard definition of what makes a great wine. Our tastes evolve. And as climate change continues to alter the wine-growing landscape, what we knew five years ago barely holds true today.

Looking ahead also means looking back, and this book is a tribute to those who laid the groundwork for our modern wine world.

I'm choosing to celebrate wine in a different way—with curiosity. Curiosity about today's new definition of what makes a great vineyard; about new regions that decades ago didn't grow vines; about historical vineyards and estates that led the way; about a beverage that's loved the world over.

As a fellow traveler, I'm glad to have you along on this journey.

To my husband, Jim, there's no one else I'd rather travel the world with than you.

OVERVIEW

AFRICA

SOUTH AFRICA

THE AMERICAS

ARGENTINA

CANADA

CHILE

MEXICO

URUGUAY

USA

OVERVIEW

ASIA

OVERVIEW

OVERVIEW

OVERVIEW

01 ESTATE VINEYARD AA BADENHORST FAMILY WINES

Kalmoesfontein farm, Jakkalsfontein Rd., Paardeberg, Malmesbury, Swartland, South Africa

TO VISIT BEFORE YOU DIE BECAUSE

This champion of old vines is one of the faces of the Swartland region.

Even though winemaker Adi Badenhorst had worked a few harvests at prestigious estates, such as Chateau Angelus and Alain Graillot in France, it was the old bush vines in the Swartland region—many at least 40 years old—that beckoned him. As any wine lover knows, older vines produce more concentrated and complex fruit. Adi's exceptional wines have made him one of the most prominent winemakers in this off-the-beaten-track region. Chenin Blanc is a specialty; 16 different parcels of the white grape are farmed on the estate. For those seeking an intimate look into life on a farm, book a stay at the on-site winemaker's cottage and enjoy walks through the vineyards.

aabadenhorst.com +27.22.1250116

02 HAMILTON RUSSELL VINEYARDS

Hemel-en-Aarde Valley, R320, Hermanus, Cape of Good Hope, South Africa

TO VISIT BEFORE YOU DIE BECAUSE

These estate vineyards were instrumental in the development of the Hemel-en-Aarde region.

Hemel-en-Aarde means "heaven-and-earth" in Afrikaans, and the two absolutely meet at Hamilton Russell Vineyards. In a quest to grow cool climate grapes, Tim Hamilton Russell purchased 170 hectares (432 acres) of undeveloped land in what is now South Africa's most southerly wine-making region. His son Anthony took over the farm in 1991 and soon focused solely on Chardonnay and Pinot Noir, which are ideally suited to the stony, clay-rich, shale-derived soil that was identified on the estate in 1994. The cooling effect of the ocean, just a few miles away, contributes to the slow and uniform development of the grapes, resulting in elegant and expressive wines. Over time, other winemakers caught on to the potential of Hemel-en-Aarde and today the region is renowned for cool-climate Chardonnay and Pinot Noir.

HamiltonRussellVineyards.com +27283123595

03 ESTATE VINEYARD – FAIRVIEW WINE ESTATE

Fairview Estate, Suid-Agter-Paarl Rd., Paarl, South Africa

TO VISIT BEFORE YOU DIE BECAUSE

You can pair homemade cheese with wines from the estate.

Fairview owns vineyards in four major regions in South Africa: Paarl, Darling, Swartland, and Stellenbosch. Paarl, the original home of the estate, is more than just vines. Third-generation winemaker Charles Back runs an entire culinary enterprise consisting of a bakery, a restaurant, a tasting room, and a goat farm, the last of which is the source of homemade cheese. Work up an appetite on one of the running or biking trails that go through the vineyards, then pair your fromage with a Shiraz, Pinotage, Viognier, Petite Sirah, or Grenache, grown on decomposed granite soils on the slopes of Paarl Mountain.

fairview.co.za 0218632450

04 KANONKOP WINE ESTATE

Kanonkop Wine Estate, R44, Elsenburg, Stellenbosch 7607, South Africa

TO VISIT BEFORE YOU DIE BECAUSE

It's the source of South Africa's best Pinotage.

Pinotage—a cross between Pinot Noir and Cinsault—was created in South Africa in 1924, and besides a few plots here and there in the world, it remains uniquely South African. It can be a finicky grape, but no one has a better handle on it than Kanonkop; about 50 percent of the Kanonkop vineyard is planted with the variety. It's a careful balance when it comes to vine vigor as vines need to struggle a bit in order to create complexity in the fruit, however, too much and there's nothing to harvest. Kanonkop's viticultural practices for the unirrigated bush vines ensure high-quality grapes every vintage. Visit the tasting room for a sip of this South African specialty.

kanonkop.co.za +27218844656

05 KLEIN CONSTANTIA ESTATE

Klein Constantia Rd., Constantia, South Africa

TO VISIT BEFORE YOU DIE BECAUSE

The wine from this estate was a favorite of Napoleon.

It takes a lot to win over a dictator, but Klein Constantia's Vin de Constance was much loved by Napoleon. The original, larger Constantia estate was founded in 1685 by Simon van der Stel but was later divided, and the 33,000 vines planted at the highest altitude on the property became what is now known as Klein Constantia. Given the vineyard's altitude and close proximity to the ocean, the cool but sunny site is an ideal place to grow Sauvignon Blanc and Muscat de Frontignan, used in the famous Vin de Constance, considered one of the greatest sweet wines in the world. Visit the estate's tasting room and restaurant to taste the liquid that softened Napoleon's heart, or book a vineyard tour.

kleinconstantia.com +27 21 794 5188

06 PAUL CLUVER FAMILY WINES

Paul Cluver Wines, De Rust Estate, N2, Grabouw, Western Cape, South Africa

TO VISIT BEFORE YOU DIE BECAUSE

The sustainably farmed vineyards thrive within South Africa's first UNESCO-recognized biosphere reserve.

The Cluver family were not only early proponents of viticulture in the cool-climate Elgin region, but they also made sustainability the forefront of their practices. Forming part of the UNESCO Kogelberg Biosphere Reserve—the first in South Africa—60 percent of the estate is marked for conservation and will never be developed. Throughout the years, the family has received numerous international awards for their farming practices. Take an e-bike tour through the vast green landscape of the property, then refuel at the restaurant, where produce comes from the estate's garden. Interested in learning more? Browse the literature at Christy's Book Nook, a book shop inside the restaurant.

cluver.com 0723336433

PAUL

07 ADRIANNA VINEYARD – CATENA ZAPATA

Cobos s/n - Luján de Cuyo (5509), Mendoza, Argentina

TO VISIT BEFORE YOU DIE BECAUSE

This high-altitude vineyard is an icon of Mendoza wine.

The Catena family pioneered high-quality wine from Mendoza, largely due to research conducted at Adrianna Vineyard. The effects of the cool climate and sun exposure at high altitudes proved grapes could ripen in a terrain previously thought inhospitable to vines. Further studies on the geology of the soils revealed its diversity; Catena vinifies anywhere from 200 to 300 different parcels from this one vineyard, each unique in character. As a result of this work, founder Nicolas Catena Zapata was given a Lifetime Achievement Award by *Wine Enthusiast Magazine* in 2021. Book a reservation for the weekly excursion to El Mirador, a lookout point at the vineyard, and see this breathtaking site for yourself.

catenazapata.com +54 261 4131100

08 ALTURA MÁXIMA VINEYARD – BODEGA COLOMÉ

Rd. 53, Km 20, Molinos (4419), Salta, Argentina

TO VISIT
BEFORE YOU DIE
BECAUSE

Altura Máxima takes grape growing to new heights.

As one of the highest-altitude vineyards in the world at 3,111 meters (10,200 feet), it's not easy growing grapes in the Altura Maxima vineyard. However, the site's unique challenges also create exciting wines: greater sun exposure and little UV protection put grapes at risk of sunburn but also produce grapes with a thick skin and a full body. Winds flowing throughout the vineyards keep the grapes cool, which helps the fruit retain acidity. Book a vineyard tour and feel as if you can touch the clouds as you walk among the vines.

 bodegacolome.com +54 9 11 2343 3633

09 DOMINIO GUALTALLARY VINEYARD – SUSANA BALBO

Cochabamba 7801, Agrelo, Luján de Cuyo, Mendoza, Argentina

TO VISIT BEFORE YOU DIE BECAUSE

This estate vineyard is owned by Argentina's first female enologist.

In 1981, Susana Balbo was the first woman in Argentina to receive an enology degree. For almost two decades she honed her craft at some of Argentina's top estates, including Catena Zapata, before finally launching her own line of wines in 1999. Dominio Gualtallary is one of her two estate vineyards, and a highly complex one at that. Three diverse layers of soils create red wines with fine-grained tannins and white wines with an earthy minerality. Entrepreneurship runs in the family: Balbo's daughter founded Osadía de Crear, one of two restaurants at Susana's Dominio Agrelo Estate where you can enjoy the wines.

susanabalbowines.com.ar +54 9 261 662 6754

10 PIEDRA INFINITA VINEYARD – ZUCCARDI

Costa Canal Uco s/n - Paraje Altamira, San Carlos, Mendoza, Argentina

TO VISIT BEFORE YOU DIE BECAUSE

12 different soil types can be found in this single vineyard.

Winemaker Sebastian Zuccardi calls Piedra Infinita a "kaleidoscope," and with 12 different soil types in this single 42-hectare (104-acre) site, it's easy to understand why. The source of these soils is the nearby Andes Mountains and the alluviums which run down the mountainside. What appears to be an endless stretch of gravel and stones covers these diverse soils and inspired the vineyard's name. Book a vineyard tour and marvel at the mosaic of soils, then taste the wines at the winery's restaurant.

zuccardiwines.com +54 2614 410000

11 ESTATE VINEYARD – FAMILIA SCHROEDER PATAGONIA

Calle 7 Norte, San Patricio del Chañar (8305) Neuquén, Patagonia, Argentina

TO VISIT BEFORE YOU DIE BECAUSE

Dinosaur bones were discovered on the property.

When Familia Schroeder in Argentina's Patagonia region started building their new winery on the top of a slope overlooking the vineyards, they unearthed a new element to their terroir: the fossils of a dinosaur. Following years of study, it was determined that this was an entirely new species. Named Panamericansaurus Schroederi after the family, a cellar was designed around the skeletal remains which rest in their original location. View the fossils in the museum-like cellar, then dine in the winery's restaurant—carnivores and herbivores welcome.

familiaschroeder.com +54 9 299 593 0187

12 ESTATE VINEYARDS – PELLER ESTATES WINERY & VINEYARD

290 John St. E, Niagara-on-the-Lake, Ontario, Canada

TO VISIT BEFORE YOU DIE BECAUSE

You can drink ice wine in a lounge made of ice.

Crafting ice wine is not for the faint of heart: grapes literally need to freeze on the vine before winemakers can press out the small amount of concentrated juice in the fruit. Peller Estate elevates sweet wine production to an art form, producing a wide range of sparkling, red, white, and sweet wines. Several immersive tasting experiences and tours take you through the estate's lineup, but the most impressive is the 10Below Icewine Lounge. Created from about 11 tonnes (25,000 pounds) of ice, it is kept at the temperature at which grapes are harvested: minus 10 degrees Celsius (50 degrees Fahrenheit).

 peller.com 1-888-673-5537

13 CLOS APALTA VINEYARD

Camino Apalta KM5, Apalta, Chile

TO VISIT BEFORE YOU DIE BECAUSE

A patchwork of plots creates Chile's version of a Bordeaux Grand Vin.

Clos Apalta, an estate from the family that created Grand Marnier Liqueur, is a study in complexity. The 60-hectare (150-acre) vineyard is laid amphitheatrically, and not only do the vines vary in age (the oldest were planted between 1915 and 1920), but the varying altitudes of the site allow many varieties such as Carménère, Merlot, Cabernet Sauvignon, and Petit Verdot to develop different expressions. Blended together, the wines from these different sites add nuance and complexity to the signature wine. Book a vineyard tour and see the magnificent spread of vineyards, then taste how these plots come together in Clos Apalta's tasting room.

ClosApalta.com 722 953350

VIÑA
SEÑA

14 SEÑA VINEYARD – SEÑA

Av. Nueva Tajamar 481, Office 503 Torre Sur, Santiago, Chile

TO VISIT BEFORE YOU DIE BECAUSE

This is the creation of two winemaking superstars.

Eduardo Chadwick of Viña Errázuriz and Robert Mondavi of his eponymous winery and Opus One are legends in their own right, but in the early 1990s they came together to create Seña. Seeing the potential of the Aconcagua Valley—similar to California decades earlier—they sought out a site that would create wines on par with Chile's northerly neighbor. They found it in Seña, a 42-hectare (104-acre) hillside vineyard. In its composite of various soils and Mediterranean climate, the grapes retain a vibrant freshness that has become a signature of Seña.

sena.cl

+569 67890200

15 VIK HOLISTIC WINERY – VIK

Millahue, San Vicente de Tagua Tagua, Chile

TO VISIT BEFORE YOU DIE BECAUSE

Numerous activities take wine appreciation to a whole new level.

The best way to appreciate wine is to engage all your senses, and at VIK winery retreat every aspect of the property is meant to ignite sensations and immerse you in the wine experience. Gaze upon the hotel—located in the center of the vineyards—which was designed by a Pritzker Architecture Prize finalist. Feel the sun on your skin during a horseback ride or bike tour through the vineyards, or listen to the rustling of the leaves on a vineyard hike. After working up an appetite, take a cooking lesson with the chef in the culinary garden and taste the earth's offerings, or marvel at the aromas in a glass of wine at The Pavilion, the on-site restaurant.

vikwine.com +56956684852

16 VIÑA DON MELCHOR

Av. Virginia Subercaseaux 210, Pirque, Region Metropolitana, Chile

TO VISIT BEFORE YOU DIE BECAUSE

Vines from Bordeaux create one of Chile's most celebrated wines.

In 1883, Don Melchor Concha y Toro brought vines from the famous French region of Bordeaux and planted them in Chile's Maipo Valley. However, it wasn't until the 1980s that these unique vines caught the eye of French vignerons visiting the region. Recognizing the potential of the terroir and using selections of these original pre-phylloxera vines, the Don Melchor wine was created. Today, the vineyard is farmed as seven distinct parcels to play to the strengths of the varieties in each subplot. Book a private Collector's Experience, which includes a tour of the vineyards and winery gardens, tastings of different parcels, and even a blending session.

donmelchor.com

+56 22 476 5000

17 BODEGA DE SANTO TOMAS

Km 94.7 Carretera Federal No. 3, Ensenada-Tecate, Mexico

TO VISIT BEFORE YOU DIE BECAUSE

You can take a carriage ride through Mexico's top vineyard.

Established in 1888, Bodega Santa Tomas set a precedent for the Mexican winemaking industry as Baja California's first winery. Large diurnal shifts between day and night, along with cooling breezes, create an ideal place for growing grapes such as Chenin Blanc, Tempranillo, and Cabernet Sauvignon. Many hands helped raise the winery's profile throughout the decades, from Dimitri Tchelistcheff (son of Andre Tchelistcheff) and Hugo D'Acosta to Laura Zamora (Mexico's first female enologist). Travel through the vineyards—and back in time—on a carriage ride, or play winemaker for a day and blend your own wine at a blending session.

santo-tomas.com +52 (664) 320 9558

18 GARZON VINEYARDS – BODEGA GARZON

Ruta 9 Km. 175, 20400 Garzón, Departamento de Maldonado, Uruguay

TO VISIT BEFORE YOU DIE BECAUSE

You'll pair your wine with the cuisine of a world-renowned chef.

Bodega Garzon is not only one of Uruguay's top destinations for wine, but for dining too. Internationally-regarded chef Francis Mallmann serves as the culinary director of the restaurant at Bodega Garzon and helped develop the restaurant's open-fire cooking concept. Pair these flame-kissed dishes with a wine (or three) from the extensive list. For those who enjoy a "wine lifestyle," the members-only club offers privileges such as wine storage in the estate's caves, and the opportunity to make wine under the guidance of Garzon's winemakers.

bodegagarzon.com (+598) 4224-1759

19 BARBOURSVILLE VINEYARDS

17655 Winery Rd., Barboursville, VA 22923, USA

TO VISIT BEFORE YOU DIE BECAUSE

This winemaker takes his knowledge from one Piedmont to another.

A native of Piedmont, Italy, Luca Paschina has been leading the winemaking at Barboursville Vineyards in the Piedmont region of Virginia almost since the estate's inception. The 364 hectares (900 acres) of land, with its varying soils and temperate weather, make a hospitable site, not just for Italian varieties such as Nebbiolo and Vermentino, but also for Cabernet Franc, Merlot, and a whole roster of other international varieties. Just like Italy's Piedmont, Virginia is also rich with history; the landmark ruins of Thomas Jefferson's mansion for Governor James Barbour are a short walk away. Wander the grounds, and you can rest assured that whichever Piedmont you are in, the wines will be delicious.

bbvwine.com 5408323824

20 BECHTHOLD VINEYARD

South of W. Woodbridge Rd. between Devries Rd. and N. Davis Rd., CA, USA

TO VISIT BEFORE YOU DIE BECAUSE

It's believed to be the world's oldest Cinsault vineyard.

While many speak of old vines in California, Bechthold in Lodi, planted in 1886, stands out for two reasons. Firstly, it is the oldest continuously farmed vineyard in Lodi (meaning it survived Prohibition) and secondly, it's regarded as the oldest Cinsault vineyard in the entire world. Originally, people believed that the vines grew Black Malvoisie, a non-premium blending grape. But DNA testing by the University of California, Davis in 2003 revealed the vines' true identity and shocked the wine industry. As a thin-skinned, aromatic, and fruity variety, Cinsault also has great aging potential. Many estates get their fruit from this famous vineyard; taste examples from Turley Wine Cellars, Poe, or Birichino.

www.visitlodi.com/blog/lodis-legendary-ancient-vine-bechthold-vineyard

21 BEDROCK VINEYARD – BEDROCK WINE CO.

414 1st Street East, Sonoma, CA 95476, USA

TO VISIT BEFORE YOU DIE BECAUSE

This Master of Wine seeks out old vines.

Master of Wine and winemaker Morgan Twain-Peterson and his business partner, Chris Cottrell, focus on reviving and preserving historic vineyards for the Bedrock label, and Bedrock Vineyard is the epitome of all they aim to achieve. Originally planted in 1854, then restored in 1888 after phylloxera ravaged the site, the centurial-plus vines have been managed by the Bedrock team since 2005. Around 30 different varieties, both old and new plantings, such as Sauvignon Blanc, thrive. There's no better setting to sample these wines than the house of General Joseph Hooker (one of the original founders of Bedrock Vineyard) in downtown Sonoma, which serves as Bedrock's tasting room.

bedrockwineco.com 707-343-1478

22 BIEN NACIDO ESTATE

4705 Santa Maria Mesa Rd., Santa Maria, CA 93454, USA

TO VISIT BEFORE YOU DIE BECAUSE

A rare transverse (east to west) mountain range influences the Pinot Noir, Chardonnay, and Syrah from this fabled vineyard.

The Bien Nacido Estate is directly in line with the cool air of the Pacific Ocean. The winds funnel through the east–west mountain range–one of only two major ranges in the U.S. with this orientation–to the interior vineyard, but the long sunny days provide ample warmth and growing time. Pinot Noir, Chardonnay, and Syrah are the specialty of this site. Along with Bien Nacido Estate's bottlings, many famous California producers also buy fruit from this historic vineyard for their wines. Taste the ocean's effect at the estate's tasting room.

biennacidoestate.com 805-318-6500

23 BERGSTRÖM VINEYARD – BERGSTRÖM WINES

8115 NE Worden Hill Rd., Dundee, OR 97115, USA

TO VISIT BEFORE YOU DIE BECAUSE

A former plum orchard now grows iconic Oregon Pinot Noir and Chardonnay.

The Bergström Vineyard had many lives—first as a plum orchard, then as a black walnut orchard, and eventually as a hazelnut orchard—before the Bergström family converted the land into a vineyard in 1999. After several years of turning the then-abandoned site into a hospitable home for grapes, the Bergströms set a new standard for Pinot Noir and Chardonnay. The bowl-shaped 5.2-hectare (13-acre) vineyard faces south so it gets optimal sunlight and grapes fully ripen. This orientation also protects it from the cold Pacific Ocean winds that blow into the valley through the Van Duzer Corridor. The Ekollon Tasting, by appointment only, is a worthwhile experience that gives you the opportunity to enjoy standard-setting Oregon wines.

bergstromwines.com 503-554-0468

24 EISELE VINEYARD

2155 Pickett Rd., Calistoga, CA 94515, USA

TO VISIT BEFORE YOU DIE BECAUSE

This iconic Napa Vineyard has been under vine since the 1880s.

When the vineyard was first planted, Riesling and Zinfandel were the varieties du jour. But almost a century later, Cabernet cemented Eisele's status as one of Napa's premier vineyards. Eisele is protected by the Palisades Mountains and westerly breezes from Chalk Hill Gap keep the grapes cool. With well-draining soils, Eisele's low-yielding vines produce fruit of exceptional complexity. Farmed organically since 1998, the vineyard went fully biodynamic in 2000. It changed hands over the decades and now the owners of Bordeaux's famed Chateau Latour are the custodians of this revered spot. For true history buffs, private meetings are held by appointment only.

eiselevineyard.com 707 942 6061

25 ESTATE SEVEN HILLS VINEYARD – L'ECOLE NO. 41

41 Lowden School Rd., Lowden, WA 99360, USA

TO VISIT BEFORE YOU DIE BECAUSE

This is one of the oldest vineyards in Walla Walla Valley.

Washington is gaining a reputation for Bordeaux-inspired wines to challenge Bordeaux itself, and Seven Hills is one of the vineyards contributing to Washington's ascent. What started as an 8-hectare (20-acre) site blossomed into 69 hectares (170 acres) when L'Ecole No. 41, in partnership with Pepper Bridge Winery and Leonetti Cellar, planted an additional 60 hectares (150 acres) in the late 1990s. With elevations between 274 -335 meters (900 to 1,100 feet) air circulates freely throughout the vines and mineral-rich soils provide nutrients to the growing plants. Learn about this important vineyard at L'Ecole's tasting room, a former schoolhouse that now educates guests about Washington's wine scene.

lecole.com 509.525.0940

26 ESTATE VINEYARDS – CHATEAU MONTELENA

1429Tubbs Lane, Calistoga, CA 94515, USA

TO VISIT BEFORE YOU DIE BECAUSE

The estate was instrumental in putting California wines on the map.

A mere four years after its establishment, Chateau Montelena shocked the world when its Chardonnay beat the French wines in the famous "Judgement of Paris" in 1976. Not only did this win bring international attention to California's wine industry, it turned Montelena into a household name. The Barrett family, who run the winery, continue the tradition of producing pinnacle California wines at their estate vineyard in Calistoga, which is planted primarily with red grapes Cabernet Sauvignon and Zinfandel. Make a reservation for a tasting at their Napa chateau and take in some history while you drink.

Montelena.com

707-94- 5105

27 ESTATE VINEYARDS – HEDGES FAMILY ESTATE

53511 N Sunset Rd., Benton City, WA 99230, USA

TO VISIT BEFORE YOU DIE BECAUSE

This estate is a leader in biodynamic farming in Washington State.

When Champagne native Anne-Marie Liégeois and Tom Hedges started exporting American wine to Sweden, neither of them could have imagined they would one day be at the forefront of the biodynamics movement in Tom's home state of Washington. Today, they are not only Demeter-certified Biodynamic in the vineyard, but in the winery too, which is a rare accolade. Popular grapes such as Cabernet Sauvignon and Merlot thrive in the sandy loam soils on Red Mountain, but experiments with Portuguese grapes such as Touriga Nacional and Souzao show a potential new future for the region. Tour the biodynamic vineyards and chateau-inspired grounds, then taste the elegant wines in the tasting room.

hedgesfamilyestate.com

509-405-5788

28 ESTATE VINEYARDS – JORDAN VINEYARD AND WINERY

1474 Alexander Valley Rd., Healdsburg, CA 95448, USA

TO VISIT BEFORE YOU DIE BECAUSE

A Bordeaux-inspired château and estate set a new standard for Sonoma half a century ago.

In their quest for Bordeaux-style elegance, Tom and Sally Jordan chose to set down roots in the pastoral land of Sonoma's Alexander Valley in 1972. On the 486 hectares (1,200 acres) of land, they built a château, one of the grandest hospitality centers in the region, and cultivated their dream Cabernets. After phylloxera ravaged the vineyards in the 1990s, the Jordans replanted as a patchwork of blocks, with only 49 hectares (120 acres) under vine. A fruit and vegetable garden, as well as an area of Tuscan olive oil trees joined the agricultural landscape, yet to this day, 364 hectares (900 acres) remain untouched. Numerous opportunities to tour the property and taste the wines abound, and newly opened suites in the château invite you to relax after a hiking excursion through the immense property.

jordanwinery.com (707) 431-5250

29 HERMANN J. WIEMER VINEYARD – HERMANN J. WIEMER

3962 NYS-14 Dundee, NY 14837, USA

TO VISIT BEFORE YOU DIE BECAUSE

This Riesling specialist contributed to the grape's prominence in the Finger Lakes.

Wiemer's family had been growing Riesling in Germany's Mosel region for over 300 years, so he knew the grape held potential in the cold climate of the Finger Lakes region in New York. The winery first started as a nursery, growing grapevines for wineries across the country in the early 1970s. A few years later, Wiemer planted his first vineyard, which was renamed Hermann J. Wiemer Vineyard after its founder in subsequent years. Today, original vines from this site still exist. Tour the winery on Seneca Lake and taste the wines that put Finger Lakes on the map.

wiemer.com (607)243-7971

30 HIRSCH VINEYARDS

57 Front Street, Healdsburg, CA, USA

TO VISIT
BEFORE YOU DIE
BECAUSE

Its proximity to a major geological fault line results in wildly diverse terroir.

The often fog-enshrouded vineyard belonging to Hirsch may look peaceful, but underneath the ground is a potential hotbed of action. Vines are located 1.6 kilometers (1 mile) away from the San Andreas fault, the point where the Pacific and North American plates meet. Movement over the years, combined with heavy rains and human and animal trekking, have resulted in a vineyard with a multitude of soils. Within a short distance the landscape can change completely, from well-draining sandstone to water-retaining clay. Due to the extremely remote location of the vineyards, they are not open to the public, but you can taste the lively Pinot Noir and Chardonnay from Hirsch at the tasting room in Healdsburg.

hirschvineyards.com 707-847-3600

31 KEUKA ESTATE VINEYARDS – DR. KONSTANTIN FRANK

9749 Middle Rd., Hammondsport, NY 14840, USA

TO VISIT BEFORE YOU DIE BECAUSE

The estate proved that vitis vinifera could grow in the Finger Lakes.

Without Dr. Konstantin Frank, the Finger Lakes would not be the renowned wine region it is today. The Ukrainian viticulturist believed that vitis vinifera could grow in the cold climate of upstate New York, and upon proving his theory, he started a vinifera revolution. He opened the door to Riesling, Gewürztraminer, and other Alpine varieties, which soon supplanted the hybrid grapes commonly found in vineyards. Vines in Frank's Keuka Lake Vineyards are now some of the oldest in the eastern United States and the Pinot Noir vines are the second oldest in all of the U.S. Tour the vineyards and cellars or enjoy a glass in the delightful wine garden.

drfrankwines.com 1-800-320-0735

32 KRONOS VINEYARD – CORISON WINERY

987 St. Helena Hwy, St. Helena, CA 94574, USA

TO VISIT BEFORE YOU DIE BECAUSE

This old-vine, phylloxera-resistant Cabernet vineyard produces one of Napa's most revered wines.

Cathy Corison is one of the most celebrated winemakers in Napa Valley and her restrained yet complex style of Cabernet has won her numerous awards. Part of her success is due to her Kronos Vineyard, a truly unique gem in this Cabernet-dominant region. Planted in 1971 on St. Georges rootstock, it proved resistant to phylloxera, the deadly vineyard disease. As a result, these over 50-year-old vines continue to produce some of the most exquisite Cabernet in all of Napa. Book an appointment at the winery where you'll sit on a crush pad and taste Cabernet just a short distance away from the vineyard.

corison.com 707-963-0826

33 KUHLKEN VINEYARD – PEDERNALES CELLARS

2916 Upper Albert Rd., Stonewall, Texas 78671, USA

TO VISIT BEFORE YOU DIE BECAUSE

It's sheer joy to taste wine from one of Texas's early wine families.

The Kuhlkens set up shop in the Bell Mountain American Viticultural Area (AVA), Texas's first AVA, in the early 1990s. At first, they just grew grapes, which were sold to a number of other wineries once word spread of their high quality. When it came time for the next generation—sibling duo David and Julie Kuhlken—to take over, they took the business in a new direction and began producing wine under the Pedernales label. The site is planted in distinct blocks with a focus on grapes such as Tempranillo, Grenache, Albariño, and Mourvedre, among others. Sample the range of Texas wines at the tasting room, or book a tour of the cellars and see how fine wine is made from grape to glass.

pedernalescellars.com

830-644-2037

34 ESTATE VINEYARD – L'AVENTURE WINERY

2815 Live Oak Rd., Paso Robles, CA 93446, USA

TO VISIT BEFORE YOU DIE BECAUSE

A maverick Bordeaux winemaker found winemaking creativity in Paso Robles.

Stephan Asseo established several châteaux in Bordeaux, but he found the French Appellation d'Origine Contrôlée (AOC) rules too stifling for the wines he wanted to make. In the mid-1990s, he uprooted his family in search of greener pastures and found them in Paso Robles. The siliceous and calcareous shale soils and large diurnal shifts (sometimes as great as 4 to 10 degrees Celsius (40 to 50 Fahrenheit) proved ideal for a wide range of varieties, particularly those from Rhône and Bordeaux. Book an appointment at the tasting room to taste these visionary cuvées.

aventurewine.com 8052271588

35 MONTE ROSSO – LOUIS M. MARTINI

254 South St., Helena Highway, St. Helena, CA 94574, USA

TO VISIT BEFORE YOU DIE BECAUSE

The site's dramatic red soils laid the groundwork for exceptional wines.

Comprised of iron-rich volcanic soils, Monte Rosso has been the source of fruit for Louis Martini's top wines for over a century. Ten varieties currently thrive on the site, including some of the first Cabernet vines, which were planted back in 1940. Old meets new at Louis Martini: the winery recently underwent major renovations and now offers new experiences, from cellar tours to tastings in a cabana.

louismartini.com

No telephone number

36 PISONI FAMILY VINEYARDS – PISONI

28275 Alta St., Gonzales, CA 93926, USA

TO VISIT BEFORE YOU DIE BECAUSE

The vines are alive with bees and butterflies at the innovative insectary.

The Pisoni family is credited with paving the way for viticulture in the Santa Lucia Highlands, and in these renowned rows of vines lies a secret garden. But this 0.8-hectare (2-acre) field is more than just plants and flora; it's an award-winning insectary and a key component in the family's sustainable farming practices. Bees, butterflies, and other flying insects flutter from flower to flower, pollinating the field while eating harmful pests that would otherwise damage the vineyards. The benefits of biodiversity ultimately produce healthier grapes. Book a tasting at the winery and find out for yourself what all the "buzz" is about.

pisonivineyards.com 8002702525

37 PYM-RAE

1100 Wall Rd., Napa, CA 94558, USA

TO VISIT BEFORE YOU DIE BECAUSE

The former Napa property of the late actor, Robin Williams, is now the wine estate of a top Bordeaux producer.

When the Tesseron family of Bordeaux's Pontet-Canet began looking for a new project, Napa, with its reputation for high-quality Bordeaux-inspired wines, called to them. After an exhaustive search, they settled upon the former estate and vineyard of actor Robin Williams. He never made wine himself but the vines, planted in 1990 in limestone and sedimentary soils, provided fruit to other estates. Using their expertise with biodynamics, the Tesserons began converting the vineyard's farming practices, even bringing in cows to act as lawnmowers. As one of Napa's newest properties, Pym-Rae showcases the new energy and focus coming to Napa.

Pym-rae.com +33556590404

38 ROOFTOP REDS

The Navy Yard, 299 Sands St., Building 275, Brooklyn, NY 11205, USA

TO VISIT BEFORE YOU DIE BECAUSE

This rooftop urban winery is defining "Brooklyn Terroir."

Located in Brooklyn's Navy Yard, Rooftop Reds is a truly urban winery. Here in the first rooftop vineyard in Brooklyn, experts from the Finger Lakes region and Cornell University helped develop the growing system that fills the skyward space. Today, 42 planters grow Bordeaux's five varieties—Cabernet Sauvignon, Merlot, Cabernet Franc, Petit Verdot, and Malbec—on the 1,375 square-meter (14,800-square foot) space. Reserve your spot at the summertime Pizza, Wine, Movie Night event where you can taste Brooklyn terroir and other New York State wines while gazing out at the city skyline.

rooftopreds.com 917 434 1288

39 RORICK HERITAGE VINEYARD – FORLORN HOPE WINES

2690 San Domingo Rd., Murphys, CA 95247, USA

TO VISIT BEFORE YOU DIE BECAUSE

The vineyard serves as both a playground and a preservation site for grape varieties.

Rorick Heritage Vineyard in California's Sierra Foothills can trace its roots back to 1844. When Matthew Rorick purchased the 30-hectare (75-acre) site in 2013, he vowed to continue the legacy of this heritage vineyard by planting and cultivating vines with a historical connection to the land. Not afraid to experiment, Rorick grows 28 lesser-known varieties such as Trousseau Noir and Pineau d'Aunis, as well as test blocks of 14 different grapes, which thrive in the high-altitude, limestone soils. Taste the range of exciting wines coming from this grape playground at Forlorn Hope's tasting room in downtown Napa.

forlornhopewines.com 209-728-8959

40 SAVOY VINEYARD – FEL

Cliff Lede Vineyards, 1473 Yountville Cross Road, Yountville, CA 94599, USA

TO VISIT BEFORE YOU DIE BECAUSE

It's the source of outstanding cool-climate Pinot Noir and Chardonnay for many wineries.

Although Anderson Valley may be under the radar when it comes to California's numerous wine regions, it's no secret that Savoy Vineyard produces some of the most exciting Pinot Noir and Chardonnay coming out of this major wine-producing state. Richard Savoy first planted the site in 1991 and acquired more land throughout the decade, bringing the total to 18 hectares (44 acres). Although today it's solely owned by FEL, many of Anderson Valley's most acclaimed wineries buy fruit and produce wines from this renowned site. Taste FEL's renditions at the tasting room of its sibling winery, Cliff Lede, and be in the know about California's most riveting Chardonnay and Pinot Noirs.

felwines.com (707) 944-8642

41 SAVOYA VINEYARD – KEN WRIGHT CELLARS

Cellars Tasting Room: 120 N Pine St. Carlton, OR 97111, USA

TO VISIT BEFORE YOU DIE BECAUSE

A native sweet onion inspired the name of one of Willamette Valley's most acclaimed vineyards.

Ken Wright is a legend in Willamette Valley, helping to create six new American Viticultural Areas (AVAs) in the region. For his own label, he sources from 13 different vineyards, but his own Savoya Vineyard is especially close to his heart. When his team started planting vines in 1999, the perfume of a sweet onion enveloped them. The source? A native onion that grew wild in the field. As a tribute to the vegetable, Wright named the site Savoya, the phonetic spelling of the Spanish word for onion. Today, the delicate and aromatic Pinot Noirs from this site continue to earn accolades. Take in the intoxicating wines for yourself at the Ken Wright Cellars tasting room.

kenwrightcellars.com 503-852-7010

42 SEVEN SPRINGS VINEYARD – EVENING LAND VINEYARDS

4180 Lone Star Rd. NW, Salem, OR 97304, USA

TO VISIT BEFORE YOU DIE BECAUSE

A former top sommelier and renowned winemaker are behind one of Oregon's most exciting vineyards.

Rajat Parr came up in the wine world as a well-respected sommelier, eventually adding author and winemaker to his resume. Winemaker Sashi Moorman's balanced and nuanced wines became north stars for Central Coast wines. After forming a bond, the duo made their mark with two California labels—Domaine de la Côte and Sandhi—before venturing north to Oregon. Upon discovering Seven Springs, they knew it was a special site for Pinot Noir and Chardonnay, and they joined the Seven Springs team in 2014. In 2021, they began the process of purchasing the vineyard from its owners, eventually taking complete stewardship of the land. See for yourself what drew these two wine-world luminaries to Oregon with an in-depth vineyard tour and tasting, or pop into the casual tasting room in downtown Dundee.

elvwines.com 503 538-4110

43 TABLAS CREEK VINEYARD

9339 Adelaida Rd., Paso Robles, CA 93446, USA

TO VISIT BEFORE YOU DIE BECAUSE

Vines from a famous Rhône estate created this vineyard and nursery.

The Paso Robles region is famous for its Rhône blends, largely due to Tablas Creek. Vines from the renowned Château de Beaucastel in southern Rhône were imported and planted in the original 49 hectares (120 acres) of rocky limestone soils, which were remarkably similar to those at the Beaucastel estate. Clones of Mourvédre, Grenache Noir, Syrah, Counoise, Roussanne, Viognier, Marsanne, Grenache Blanc, and Picpoul Blanc can be credited with starting the Rhône "revolution." In later years, Tablas Creek built its own nursery, propagated the remainder of the 13 permitted Rhône varieties, and now sells these varieties to growers interested in cultivating the essence of the Rhône in their vineyard. Fantasize about how you'd create your own version of Rhône wines on a tour of the nursery and the estate's Regenerative Organic Certified vineyard.

tablascreek.com

8052371231

44 TO KALON VINEYARD – ROBERT MONDAVI WINERY

7801 Saint Helena Highway, Oakville, CA 94562, USA

TO VISIT BEFORE YOU DIE BECAUSE

This valley floor vineyard is Napa's "highest beauty."

The To Kalon Vineyard, meaning "highest beauty" in Greek, can trace its history to the 1800s. Even then it was highly regarded; wines won awards at fairs as far away as Paris. Although phylloxera ravaged many of the vines at the turn of the century, a replanting turned the site into one of Napa's "Grand Crus." Robert Mondavi, who founded the iconic Robert Mondavi Winery and started working on the site in 1966, touted its ability to produce wines with beautiful texture, complexity, and age-ability. At Robert Mondavi Winery, there are numerous opportunities to explore the mystique of To Kalon, from outdoor tastings to special pairing lunches and dinners, as well as events and music concerts.

robertmondaviwinery.com (888) 766-6328

45 VAN DER KAMP VINEYARD

6360 Sonoma Mountain Rd., Santa Rosa, CA 95404, USA

TO VISIT BEFORE YOU DIE BECAUSE

The site of old-vine Pinot Noir is believed to be where the world began.

On top of Sonoma Mountain sits one of California's oldest Pinot Noir vineyards. Vines, at 458 meters (1,500 feet) in elevation and dating back to 1952, cover the picturesque site. Indigenous people believed this was where the earth began. At the very least, beautiful wines get their start here. The vineyard has numerous soils, many containing elements of volcanic material, which give unique expressions to the fruit. The van der Kamp family produce their own wines and provide fruit to a number of top Sonoma estates. Open by appointment only, book a time for a picnic atop this scenic site and ponder the origins of great wine.

vanderkampvineyard.com 707-528-0894

46 YAVAPAI COLLEGE ESTATE VINEYARD – SOUTHWEST WINE CENTER

601 Black Hills Drive, Clarkdale, AZ 86324, USA

TO VISIT BEFORE YOU DIE BECAUSE

A rock star helped fund this enology school.

Cacti and vines co-exist in this desert-climate vineyard, a training ground for hopeful winemakers in the Southwest U.S. The project kicked off when Arizona winemaker and musician Maynard James Keenan (Tool, A Perfect Circle, Puscifer) donated a 0.4-hectare (1 acre) block of Negroamaro to the fledgling project in 2009. Since then, the education center has grown to include several more blocks of grape varieties as well as a working winery. Taste the students' graduation projects at the tasting room—and don't worry, there won't be a quiz at the end.

southwestwinecenter.com 9286346566

47 WÖLFFER ESTATE VINEYARD

139 Sagg Rd., Sagaponack, NY 11962, USA

TO VISIT BEFORE YOU DIE BECAUSE

Former potato fields now produce exquisite wines in New York's vacation destination, The Hamptons.

Nestled among the mansions of the rich and famous sits Wölffer Estate, home to delicious and well-crafted wines and ciders. The Atlantic Ocean and its cooling breezes play a big part in creating the maritime climate on the South Fork of Long Island, and grapes develop beautiful acidity. Given the vacation-destination ethos of the Hamptons, it's no surprise that rosé is one of Wölffer's most popular wines. In addition to touring the vineyard and winery, grab a blanket and a bottle and enjoy live music al fresco on a warm summer night on the property.

wolffer.com 631.537.5106

48 ADONG, SHUORI, SINONG, AND XIDANG – MOËT HENNESSY SHANGRI-LA

Adong, Deqin, Shangri-la, Yunnan, China

TO VISIT BEFORE YOU DIE BECAUSE

China's vineyards bring new meaning to mountain wines.

The Himalayas serve as a dramatic backdrop to Moët Hennessy's first Chinese winery, a project which started in 2008. In the remote area of Shangri-La, along the banks of the Mekong River, vines are cultivated in four different villages (Adong, Shuori, Sinong and Xidang), each with its own expression of terroir. Ranging from 2,100 to 2,800 meters (6,890 to 9.186 feet) above sea level, the vineyards have different orientations, soils, and climates that produce fruit with very distinct profiles. About 120 families farm and hand-harvest the grapes, all of which are grown organically. Blended together, these sites create Ao Yun Grand Cru.

aoyun-wine.com 67546389

49 ESTATE VINEYARD – TSINANDALI ESTATE

Tsinandali Village, Kakheti, Georgia

TO VISIT BEFORE YOU DIE BECAUSE

Ancient and modern winemaking meet at this Georgian estate vineyard.

With a winemaking history dating back 8,000 years, the country of Georgia is one of the earliest places for wine production in the world. At Tsinandali, the 40-hectare (99-acre) vineyard grows 11 different indigenous Georgian grapes, but the international variety Pinot Noir also makes an appearance. Traditional *qvevri*—large egg-shaped clay pots—and state-of-the-art equipment are used to produce the estate's wide range of wines. Book a tour of the vineyard for an up-close look at Georgia's wine history, then feast on a traditional Georgian meal among the vines.

tsinandaliestate.ge +995 568 99 99 59

50 CHATEAU MUKHRANI VINEYARDS – CHATEAU MUKHRANI

Chateau Mukhrani, 3309 Georgia Mtskheta, Georgia

TO VISIT BEFORE YOU DIE BECAUSE

A descendant of the Royal Family brought modern winemaking to the "Cradle of Wine."

After visiting France in the late 1800s, Ivane Mukhranbatoni, a descendant of the Bagrationi Royal Family, brought French winemaking techniques back to his native country of Georgia and its centuries-long history of viti-culture. Today, this cross-culture influence can be seen in the organically-farmed vineyard; seven varieties of indigenous Georgian grapes and seven varieties of international grapes coexist. It's easy to feel like royalty yourself when visiting the estate. Tour the grand castle, explore the underground cellars or ride a horse through the vineyards.

chateaumukhrani.com

+995 595 99 13 14

51 SULA VINEYARDS

Govardhan Village, Off Gangapur-Savargaon Rd., Nashik, Maharashtra, India

TO VISIT BEFORE YOU DIE BECAUSE

This prominent Indian winery is also an ideal getaway.

When Rajeev Samant planted the first vines in the region of Nashik in 1999, little did he know this experiment would lead to India's largest wine-growing region. Today, 80 percent of India's wine is produced in Nashik. Sula's plantings include Cabernet Sauvignon, Shiraz, and Zinfandel, along with white varieties such as Chardonnay and Viognier. The winery itself has grown into a premier destination and an array of activities entice visitors, from vineyard tours and tastings to bike rides and vineyard picnics.

sulavineyards.com 99700 90010

52 ESTATE VINEYARDS – VITKIN WINERY

Turtle Bridge, Alexander Stream, Kfar Vitkin, Israel

TO VISIT BEFORE YOU DIE BECAUSE

This Israeli winery champions lesser-known grape varieties.

As residents of Kfar Vitkin (a cooperative farming village), agriculture was second nature to Vitkin Winery's founders. Instilled with the ethos of hard work, they not only started a winery but chose to cultivate varieties beyond the popular Chardonnay and Cabernet Sauvignon. Everything from Riesling to Carignan and Grenache Blanc grows in the vineyards, and new trials of grapes, such as Marselan, demonstrate Vitkin's forward-thinking mentality. Ponder your own next big move as you snack on small bites paired with wines at the visitor center.

vitkin-winery.co.il

+972 9-866-3505

53 YIRON VINEYARD – GALIL MOUNTAIN

Kibbutz Yiron, Upper Galilee, Israel

TO VISIT BEFORE YOU DIE BECAUSE

A joint project between Golan Heights Winery and Yiron Kibbutz is elevating modern Israeli winemaking.

Galil Mountain Winery, located in Upper Galilee, draws on the modern winemaking knowledge of the Golan Heights Winery and the traditional agricultural focus of Kibbutz Yiron (a community settlement) for its inventive wines. The largest of Galil Mountain Winery's seven estate vineyards, Yiron Vineyard is planted with numerous French and Italian varieties, such as Barbera, Syrah, Cabernet Sauvignon, and Sauvignon Blanc. The range of soil types, various expositions, and large diurnal shifts in the summer growing months allow for this diversity of grapes. Gaze at the ancient land while sipping wine in the sleek and airy tasting room.

galilmountain.co.il 972-4-6868748

54 MIYA-NO-ME – LUMIÈRE WINERY

624 Minaminoro, Ichinomiya-cho, Fuefuki-shi,
Yamanashi 405-0052, Japan

TO VISIT
BEFORE YOU DIE
BECAUSE

The winery was founded by the family that once served as chief priests of Ōmiya Shrine.

Lumière, founded in 1885, is one of Yamanashi Prefecture's most respected wineries. Run by a fifth generation, the "Miya-no-mae" vineyard is located in front of the Ōmiya Shrine, which is adjacent to the modern-day winery. Due to Japan's strict regulations regarding land ownership, the vineyard is only 0.57 hectares (1.4 acres), but fruit is sourced from a total of 30 hectares (74 acres) throughout Yamanashi and Nagano prefectures. Grapes from this site goes into Lumière's top series "Hikar" and its flagship wine, Château Lumière. Dine at Zelkova, the on-site French-inspired restaurant, as you gaze out at the vineyard.

lumiere.jp/english

+81-55-347-0207

55 UTSUGIDA VINEYARD – MGVS

601-17 Todoroki, Katsunuma-cho, Koshu-shi, Yamanashi 409-1315, Japan

TO VISIT BEFORE YOU DIE BECAUSE

A highly scientific approach to Koshu and Muscat Bailey A results in stunning wines.

MGVs looks partly like a winery and partly like a science lab. The owner, Hiroshi Matsuzaka, brings his experience operating a semiconductor manufacturing facility to winemaking. In the Utsugida estate vineyard, grape bunches of Koshu might don little rain caps to protect them from hail and rain during the rainy season. Numbers and letters on labels read like a table of elements and correlate to wine-growing areas, vinification methods, and more. But as any wine lover knows, terroir is a major factor in a grape's character, and degraded granite and volcanic ash from Mounts Daibosatsu and Kyodo imbue the Koshu grape with a vibrant, mineral quality. Conduct your own wine-tasting experiment in the sleek tasting room.

 mgvs.jp/eng 81-55-344-6030

56 BEKAA VALLEY – CHÂTEAU MUSAR

Chateau Musar, Ghazir, Lebanon

TO VISIT BEFORE YOU DIE BECAUSE

These age-worthy wines come from one of the world's oldest agricultural sites.

Château Musar's vineyards reside in the Bekaa Valley, nearly two hours away from the winery, but the commute is worth it: 6,000 years of winemaking history prove the region is an ideal spot for grape cultivation. Because terroir like this has stories to tell, Musar doesn't release the wines—both red and white—until seven years after harvest to allow them to evolve and start showing their true potential. Take the time to explore the winery on a tour, then taste these expressive wines in the tasting room.

chateaumusar.com +961 9 925 056

57 DOMAINE DES TOURELLES

Chtaura, Bekaa Valley, Lebanon

TO VISIT BEFORE YOU DIE BECAUSE

Here you'll find old-vine Cinsault from a historic estate.

With a warm, dry climate, an altitude of 1,050 meters (3,445 feet) above sea level, and 240 days per year (on average) of sunshine, the Bekaa Valley is an ideal place to cultivate grapes and to farm organically. Family-run Domaine des Tourelles, founded in 1868, dry farms the old bush vines. Without irrigation, vines learn to make do with their surroundings, and the practice results in complex, concentrated fruit. The majority of the vineyards are made up of 70-year-old bush vines. The Cinsault, in particular, is a standout. Enjoy wines in the winery's garden as you taste the estate's numerous offerings.

domainedestourelles.com +9613805925

58 GRANMONTE VINEYARD AND WINERY

52 Moo 9 Phayayen, Pakchong, Nakhon Ratchasima 30320, Thailand

TO VISIT BEFORE YOU DIE BECAUSE

You can explore the property by bike at this Thai vineyard.

In GranMonte's pocket of Thailand, the climate defies the country's tropical reputation. Thanks to the Khao Yai mountain range, GranMonte's estate vineyard remains protected from the tropical storms and monsoons that would otherwise be detrimental to wine growing. The combination of lower annual rainfall than the rest of Thailand and less humidity enables the estate to grow a range of varieties, such as Syrah, Cabernet Sauvignon, Grenache, Chenin Blanc, Viognier, and more. Tour the tropical vineyard by bike, then pair GranMonte's wines with lunch at the on-site restaurant.

www.granmonte.com +66 92 806 7755

59 ESTATE VINEYARD – YACOUBIAN-HOBBS ESTATE

Aghavnadzor, Vayots Dzor, Armenia

TO VISIT BEFORE YOU DIE BECAUSE

An internationally renowned winemaker brings his skills to this Armenian vineyard.

Paul Hobbs makes wines in places as far flung as Argentina, Spain, France, and California, and with Yacoubian-Hobbs, he aims to showcase indigenous Armenian varieties. The vineyard resides in the region of Vayots Dzor, home to the Areni-1 cave. Considered to be the world's oldest known winery, it dates back to 4100 BC. The red Areni grape thrives in high-altitude volcanic soils such as those found at the Yacoubian-Hobbs estate vineyard. A handful of white varieties completes the vision for this Armenian endeavor.

 yacoubianhobbs.com 707-824-9879

60 ACHLEITEN – DOMÄNE WACHAU

A-3601, Dürnstein 107, Austria

TO VISIT BEFORE YOU DIE BECAUSE

This incredibly steep and diverse vineyard is the source of many of the Wachau's greatest wines.

No single producer can lay claim to Achleiten, and that's a good thing, since that gives more the opportunity to express the potential of this vineyard. Steeply sloped, the Danube runs along the base while a forest crowns the top. A stone wall surrounds the vines and runs upward, and at about 210 meters (689 feet) above sea level, a terracing system begins. These stone ledges make the highest vines accessible while also preventing erosion. As one of the latest-ripening vineyards in all of the Wachau, wines showcase an incredible amount of power, as well as a smoky, flinty mineral quality. Gaze upward at this majestic site on a boat tour from Domäne Wachau, or reach new heights of wine appreciation with a tasting at the tavern.

domaene-wachau.at +432711371

61 ESTATE VINEYARD – GUT OGGAU – FAMILIE TSCHEPPE-ESELBÖCK

Hauptstrasse 31, 7063 Oggau, Austria

TO VISIT BEFORE YOU DIE BECAUSE

An abandoned vineyard became the source of biodynamic wines from one of Austria's most celebrated estates.

When Stephanie and Eduard Tscheppe-Eselböck took over the property that now houses their winery, major renovations needed to be done to the 17th-century structure and vineyard. Since the vines had been abandoned for 20 years, they fortunately had time to rest and reset, and any residual effects of chemicals and pesticides had disappeared. The couple revitalized the site and converted it to biodynamic farming. Today, Gut Oggau is Demeter-certified. During the spring and summer, make a reservation at the winery's *hueriger* (tavern) and enjoy a traditional feast paired with Gut Oggau's personality-driven wines.

 gutoggau.com +43 664 2069298

62 RIED ULM - NUSSBERG – WEINGUT WIENINGER

Stammersdorfer Strasse 31, 1210, Vienna, Austria

TO VISIT
BEFORE YOU DIE
BECAUSE

These vineyards reside in the city of Vienna.

With over 600 hectares (1,482 acres) of vines Vienna is the only European city to be considered a wine-growing appellation. One of the most exciting of these urban vintners is Fritz Wieninger, and the Ried Ulm - Nussberg vineyard is the source of world-class wines. A city skyline isn't your typical vineyard view, but that's what you'll see when you stand on the site's marine sediment and limestone soils. From this vineyard comes a top example of Wieninger's specialty: Gemischter Satz, essentially a field blend of several varieties. Book a reservation at the estate's *heuriger* (tavern) or seasonal *buschenschank*, located in the middle of a vineyard, for a true country-meets-city experience.

www.wieninger.at +43 1 2901012

63 TURNER VINEYARD – HEIDI SCHRÖCK & SÖHNE

Rathausplatz 8, A-7071 Rust, Austria

TO VISIT BEFORE YOU DIE BECAUSE

The leader of Cercle Ruster Ausbruch revitalized a historic Austrian wine.

Before phylloxera ravaged vineyards in Europe in the late 19th century, Austria produced a wine called Ruster Ausbruch that was on par with the world's best sweet wines. It didn't make a comeback, however, until a group of winemakers—led by Heidi Schröck—called Cercle Ruster Ausbruch was formed. Schröck already had a penchant for creating change: she reintroduced Hungarian varieties such as Furmint and Muscat Lunel to Burgenland. Due to the challenging conditions required to create the sweet wine—grapes must be infected by the "noble rot" botrytis in order for sugars to concentrate—it took a lot of training and campaigning to bring it back to prominence. Thanks to Schröck, it received DAC status in 2020. Book a bike tour through the vineyards, then taste the sweetness of victory at the winery's tasting room.

heidischroeckundsoehne.com +43 2685 229

© Wijndomein Aldeneyck

64 ESTATE VINEYARDS – WIJNDOMEIN ALDENEYCK

Hamontweg 81, 3680 Maaseik, Belgium

TO VISIT
BEFORE YOU DIE
BECAUSE

You'll forget all about Belgian beer with one sip of wine from this Belgian estate.

Belgium may be closely associated with beer, but there's a lively wine-making scene taking place there too. Wijndomein Aldeneyck, near the city of Maaseik, cultivates vines in the Limburg Meuse Valley. Vineyards, located on southeastern slopes, enjoy the sunny exposure, while pebble and gravelly soils provide minerals to the vines. Pinot Blanc, Pinot Gris, Pinot Noir, Chardonnay, and Riesling thrive and create the backbone of the estate's still and sparkling wines. Book a tour of the vineyard and cellar, and during warmer months sip Belgian's finest wines on the outdoor terrace.

© Wijndomein Aldeneyck

wijndomein-aldeneyck.be +3289564102

65 BAS DE BAS SELF-SUSTAINABLE VINEYARD ESTATE – BIBICH WINERY

Vaćani BB, 22222, Skradin, Croatia

TO VISIT
BEFORE YOU DIE
BECAUSE

It's a delight to sip poolside wine while gazing out at Croatian vineyards.

Wine vacation comes full circle as you relax at one of BIBICh Winery's wine villas, drinking the signature Bas de Bas wine just steps away from the vines themselves. The family-run estate cultivates Syrah, Merlot, and Sauvignon Blanc on its 6 hectares (15 acres) of vineyards. Renovated, solar-powered villas, formerly belonging to the estate manager and staff, invite guests to unwind. Whether swimming in the estate's swimming pool or hiking through the vineyards, there are plenty of ways to appreciate this Croatian retreat.

bibich.co/basdebas +385913235729

66 NJORD VINGAARD

Tjørnedevej 5, 4300 Holbæk, Denmark

TO VISIT BEFORE YOU DIE BECAUSE

One of Denmark's few red wines is made from a unique mutation of Pinot Noir.

Pinot Noir may be the grape behind some of the world's best wines, but have you ever tasted Pinot Noir Précoce? As an early mutation of Pinot Noir (writings from 1600 note its existence), it's proving to be a red variety that thrives in Denmark. At Njord Vingaard, the early-ripening grape, with its thicker skins, and slightly lower acid and tannin profile, is a good match for the country's long hours of sunshine in the early part of the growing season and maritime climate. While the winemaking trend in Denmark is predominately white wines, book an appointment at Njord to discover how a historic red grape is creating something completely new.

njordvingaard.dk

+45 31904180

67 GUSBOURNE VINEYARD – GUSBOURNE ESTATE

Gusbourne Estate, Kenardington Rd., Appledore, Ashford, Kent TN26 2BE, UK

TO VISIT BEFORE YOU DIE BECAUSE

History is unearthed at this English sparkling wine estate.

When the team at Gusbourne began converting 61 hectares (150 acres) of land in Appledore, Kent, into vineyards, they uncovered much more than just soil. Coins dating back to around 1051, a 16th-century ring, and ammunition rounds from WWII, just to name a few items, turned up on the site. Many of these tell the centuries-old history of Appledore as a port city and are on display at Gusbourne's estate. While you may not find these treasures on your vineyard walking tour, you can, however, view the grape bunches that make Gusbourne's sparkling wines worth their weight in gold.

gusbourne.com 01233 884680

68 NYETIMBER MANOR VINEYARD – NYETIMBER ESTATE

Gay Street, West Chiltington RH20 2HH, UK

TO VISIT BEFORE YOU DIE BECAUSE

This vineyard led the way for Champagne-style sparkling wine in England.

The marine subsoil stemming from the Paris Basin that runs through Champagne also makes its way to England and is a big contributor to England's sparkling wine revolution. Nyetimber was at the forefront of the moment: in 1988 the estate planted Champagne's three key grapes—Pinot Noir, Chardonnay, and Pinot Meunier—on what was to become the estate vineyard. With these three key varieties, they followed Champagne's method of creating sparkling wine, which is by second fermentation in bottle. When Nyetimber started winning awards less than a decade later, a sparkling wine passion was ignited. Spark your love for English bubbly at one of the estate's Open Weekends during the summer or at one of the chef's pairing dinners.

nyetimber.com +44 (0) 1798 813989

69 CHAMPAGNE PASCAL AGRAPART

57 Avenue Jean Jaurès, 51190 Avize, France

TO VISIT BEFORE YOU DIE BECAUSE

Here you can enjoy Grand Cru Champagne from an esteemed grower-producer.

Many Champagne houses buy fruit from grapes growers, but Champagnes produced and bottled by the farmers themselves (a category known as *Récolant Manipulant*) are becoming well-known among Champagne fans. Some of the most exciting come from fourth-generation Pascal Agrapart in the village of Avize, in particular his vintage Champagnes. Chalky limestone forms the subsoil but different levels of clay topsoil along the slope create distinct Champagnes; Minéral at the bottom contains less clay; L'Avizoise comes from the clay-rich soils at the top of the slope; and Vénus, derived from a single vineyard of old vines, lies at a favorable position mid-slope. When the house's new bed and breakfast opens soon, you can start your morning with a proper glass of Champagne.

champagne-agrapart.com +33 3 26 57 51 38

70 CHÂTEAU ANGÉLUS

Château Angélus, 33330 Saint-Emilion, France

TO VISIT
BEFORE YOU DIE
BECAUSE

This superstar estate rings out with love for Cabernet Franc.

In Bordeaux's Right Bank, Merlot is the dominating force, but for Château Angélus, Cabernet Franc does most of the heavy lifting for its high-ranking cuvée. Located in a south-facing amphitheater, the vineyard of this esteemed estate contains sandy-limestone soils at lower elevations (where Cabernet Franc thrives) and heavier clay soils further up the slope (a hospitable home for Merlot). It was said that from this amphitheater, the sound of the Angélus bell could clearly be heard during its thrice-daily ringing, and guided the daily rhythms of the workers. In homage to this, the image of a bell graces the wine's label. Taste the wines at Michelin-starred restaurants Logis de la Cadène in Saint-Emilion or Le Gabriel in Bordeaux.

angelus.com 0557247139

71 CHÂTEAU DE SELLE – DOMAINES OTT

5093 Route de Flayosc, 83460 Taradeau, France

TO VISIT
BEFORE YOU DIE
BECAUSE

The site of former lavender fields became the source of quintessential Provence rosé.

Provence is the birthplace of rosé, and no estate is more synonymous with the elegant pink wine than Domaines Ott. Château de Selle was the first of the three properties purchased by Marcel Ott. Seeing the potential in the inland slopes covered with olive groves and lavender fields, he set about planting a range of varieties—Cabernet Sauvignon, Syrah, Cinsault, and Grenache—which form the backbone of Domaines Ott's famous wine. Find eternal summer at a tasting in the winery's tasting room.

 domaines-ott.com/en +33 4 94 47 57 57

72 CHÂTEAU LATOUR

Château Latour, Saint Lambert, 33250 Pauillac, France

TO VISIT BEFORE YOU DIE BECAUSE

This estate is one of the original "First Growths" from the Bordeaux Classification of 1855.

In 1855, one of the most important events in the history of fine wine took place: the Bordeaux Classification system. Created by Napoleon III, it ranked all the Bordeaux chateaux at the time, creating a hierarchy that continues to influence the wine world. The grapes for the Grand Vin de Château Latour come exclusively from old vines (with an average age of 60 years) planted in l'Enclos—the estate's historic 47-hectare (116-acre) hilltop vineyard. Clayey gravel, ideal for Cabernet, makes up the vineyard's interior, while a gravely sand, home to the Merlot vines, occupies the perimeter. Latour is a true original.

chateau-latour.com

+33556731980

EXPERIMENTATION
Saury
Saury
Châteauroux

73 CHÂTEAU PONTET-CANET

Château Pontet-Canet, 33250 Pauillac, France

TO VISIT BEFORE YOU DIE BECAUSE

This well-regarded Bordeaux estate was an early proponent of biodynamics.

Although nowadays much of Bordeaux embraces sustainable farming practices, Pontet-Canet identified the need early on. In the mid-1990s, the estate started farming organically, and in 2004 it trialed biodynamic farming on 14 hectares (35 acres) of land. When results proved successful, the owners began the process of converting the entire 81-hectare (200-acre) vineyard to biodynamics. As they were one of the first Classified Growth chateaux to farm in this way, their action was revolutionary. In 2010 the Chateau became certified in organic and biodynamic agriculture and is one of the few estates to hold this double certification.

pontet-canet.com +33556590404

74 CLOS D'ORA – GÉRARD BERTRAND

Château l'Hospitalet, Route de Narbonne-Plage, 11100 Narbonne, France

TO VISIT BEFORE YOU DIE BECAUSE

This estate is owned by a former rugby player who is now the biggest biodynamic producer in the Languedoc.

Former rugby player Gérard Bertrand never set out to be a vigneron, but after the death of his father, he took over the family estate and started his own legacy by buying numerous estates and, most importantly, converting them all to biodynamic farming. Clos D'Ora, a delineated portion of the Laville Bertrou estate that sits on a geological fault of limestone and marl soils, is a jewel in the wine empire. After obtaining the estate, Bertrand planted Syrah and Mourvèdre to complement the old vines of Carignan and Grenache that already existed on the site. Blended together, they became the signature Clos D'Ora wine. Feel a seismic shift when you taste the wine for yourself at Chateau l'Hospitalet, Bertrand's recently renovated wine estate, resort and spa.

 gerard-bertrand.com +33 468452850

75 CLOS DE LA ROCHE GRAND CRU – DOMAINE DUJAC

7 Rue de la Bussière, 21200 Morey-Saint-Denis, France

TO VISIT BEFORE YOU DIE BECAUSE

This Grand Cru vineyard produces some of the most celebrated Pinot Noir in all of Burgundy.

With its limestone and clay soils and great sun exposure, Pinot Noir from Clos de la Roche is often identified by its cherry and truffle character. Several producers own vines on the 8 *climats* (plots) that exist within the vineyards, one of the most acclaimed being Domaine Dujac. The estate, founded by legendary winemaker Jacques Seysses, works with five of the plots and owns a total of 1.46 hectares (3.63 acres) out of the 16.91 hectares (41.8 acres) that make up the whole of Clos de la Roche. Today, Jacques's sons Jeremy and Alec, and Jeremy's wife, Diana, continue the work started by the previous generation and the wines remain some of the most sought after in the world.

dujac.com

+33 3 80 34 0100

76 CLOS DU BOURG – DOMAINE HUET

11 rue de la Croix Buisée, 37210 Vouvray, France

TO VISIT BEFORE YOU DIE BECAUSE

It has been called the greatest vineyard in all of Vouvray.

Founded in 1928, Domaine Huet carries two reputations: one of being Vouvray's greatest estate, and the other as a leader in biodynamic farming. Looking for a quieter way of life, Parisian bistro owner Victor Huet moved to the Loire and purchased his first vineyard. Over time the family acquired two more sites, one of which was Clos du Bourg. Located above a church, the vineyard itself is surrounded by a 9th-century stone wall. With shallow clay and limestone soils, it produces powerful and age-worthy Chenin Blanc. The Huet family was an early proponent of biodynamic farming, and credit the complexity and longevity of their wines to their practices. Book an appointment to spend time in the tasting room contemplating these revered wines.

domainehuet.com

+33 2 47 52 78 87

77 CLOS MONTMARTRE

Vignes du Clos Montmartre, Rue des Saules 75018 Paris

TO VISIT
BEFORE YOU DIE
BECAUSE

This is a true urban vineyard in the heart of Paris.

A grape oasis in the center of Montmartre, Clos Montmartre was planted in 1933 as a way to save the city from urbanization, and it has become a treasure of the 18th arrondissement. Along with Pinot Noir and Gamay, a range of varieties commonly found throughout France's wine-growing regions thrive in this garden of vines. Celebrate the newest vintage of this Parisian wine the second weekend of every October at the Montmartre Harvest Festival.

www.comitedesfetesdemontmartre.com 0684156095

78 CLOS SAINT HILAIRE – CHAMPAGNE BILLECART-SALMON

40 Rue Carnot, 51160 Mareuil-sur-Aÿ, France

TO VISIT BEFORE YOU DIE BECAUSE

A return to the past makes this single parcel a remarkable site for Pinot Noir.

Clos Saint Hilaire is a small, 1-hectare (2.47-acre) parcel within a larger vineyard, and the Billecart-Salmon family identified these few precious rows as the source of extraordinary Pinot Noir. To truly bring out the essence of the fruit from this tiny clos, the family returned to ancestral farming methods: only workhorses, sheep, and humans tend to the vines. As a vintage Champagne, it is solely made in years when conditions are perfect, and only 3,500 to 5,500 bottles are produced. Book an appointment for a tour and tasting, and be blown away by the great Champagne that comes from this minuscule plot.

champagne-billecart.fr

No telephone number

79 CLOS ST. HUNE – TRIMBACH

15 Route de Bergheim, 68150 Ribeauvillé, France

TO VISIT BEFORE YOU DIE BECAUSE

Thirteen generations have tended to this vineyard-within-a-vineyard.

At Trimbach, it's a true family affair: 13 generations in fact. But when one has a site like Clos St. Hune to work with, it's no wonder the family wants to stay in the business. The Clos is an exclusive 1.67-hectare (4.12-acre) site within the Rosacker Grand Cru vineyard and has been owned by the family for over 200 years. Vines are between 50 to 70 years old and grown on Muschelkalk, a soil containing fossilized seashells. Taste the wine at the Trimbach tasting room and feel like part of the family.

trimbach.fr

+33(0)389736030

80 THE CLOS SAINT URBAIN, RANGEN DE THANN GRAND CRU – DOMAINE ZIND HUMBRECHT

4 Route de Colmar, 68230 Turckheim, France

TO VISIT BEFORE YOU DIE BECAUSE

Here you can pay homage to a saint beloved by winemakers.

Saint Urbain was particularly beloved by the residents of Thann, and in the 15th century a chapel was erected in his honor in what is now known as the Rangen de Thann Grand Cru vineyard. Every year, wine producers and innkeepers formed a procession and visit the chapel, which has been restored several times in its history. Today, vines surround the chapel, mostly Pinot Gris, Riesling, and Gewürztraminer. Book a weekday tour of Zind-Humbrecht's cellars for a taste of this saintly wine.

zindhumbrecht.fr/en

+33 3 89 27 02 05

81 DOMAINE DE LA RÉSERVE D'O

Rue du Château, 34150 Arboras, France

TO VISIT
BEFORE YOU DIE
BECAUSE

This windswept, biodynamically-farmed vineyard produces some of the most exciting wines from the Languedoc.

When Marie and Frédéric Chauffray started their vineyard in 2004, they saw it as a holistic ecosystem, and even the name expresses their commitment to the idea of a nature reserve. From the get-go, they farmed biodynamically and continually improve their methods of low-intervention winemaking. It helps that the vineyard is located 400 meters (1,310 feet) above sea level in the foothills of Plateau du Larzac where winds continually sweep the vineyards clean. Join the Chauffreys on a stroll of the vineyard and be amazed at their stunning wines.

lareservedo.fr +33 6 76 04 03 88

82 DOMAINE TEMPIER – SAS PEYRAUD

1082 chemin des Fanges Peyraud, 83330 Le Castellet, France

TO VISIT BEFORE YOU DIE BECAUSE

Drink wine from the "Godfather of Bandol" and champion of Mourvèdre.

When Lucien Peyraud and Lucie "Lulu" Tempier married in 1936, they were gifted a farm by Lulu's father. Lucien, an aspiring winemaker, became smitten with a bottle of wine from the estate before its vines were decimated by phylloxera. His research brought him to the conclusion that Mourvedre would create stunning, age-worthy wines and he tirelessly petitioned the French government to grant Bandol its own AOC status, based on the quality of the wines being produced. He succeeded and today all Bandol AOC wines must contain at least 50 percent Mourvèdre. Visit the Tempier tasting room and taste the wines from the vineyards where it all started.

 domainetempier.com 0494987021

83 E. GUIGAL

Château d'Ampuis, 69420 Ampus, France

TO VISIT
BEFORE YOU DIE
BECAUSE

The "roasted slopes" of the northern Rhône produce some of the most sought-after red wines.

The steep, terraced slopes of Côte Rôtie have been under vine since Roman times, but it wasn't until Etienne Guigal arrived that the region, felled at different times by wars and phylloxera, began to gain respect. Côte Rôtie is affectionally divided into Côte Blonde and Côte Brune, which refer to two soil types: dark brown mica schist *(brune)* and pale yellow gneiss *(blonde)*. Syrah is the focus, but the estate cultivates a modest amount of Viognier for blending purposes. Today, these tiny-production wines are highly coveted. Explore Guigal's importance in the region's history with a visit to the museum at Le Caveau du Château where you can taste your way through the region's best.

guigal.com

+33 (0)4 74 56 10 22

84 ESTATE VINEYARDS – CHÂTEAU DE BEAUCASTEL

Chemin de Beaucastel, 84350 Courthezon, France

TO VISIT BEFORE YOU DIE BECAUSE

The famous millennia-old *galet* stones of Châteauneuf-du-Pape are a sight to behold.

When one looks at the large round stones that cover Châteauneuf-du-Pape vineyards, it doesn't seem an obvious place to grow vines. But those smooth pebbles, called *galets* are a vine's secret weapon; they gather heat during the day and slowly release it at night, helping to regulate temperatures in the vineyards. And there's a lot to protect: Châteauneuf-du-Pape allows up to 13 different grape varieties to be blended in the appellation's cuvées. Beaucastel, one of the region's preeminent estates, has perfected the art of blending—a complex and detailed process considering they farm over 100 hectares (247 acres) of vineyards in total.

beaucastel.com +33 (0) 4 90 70 41 00

85 ESTATE VINEYARD – DOMAINE AUX MOINES

5 Chemin de la Roche aux Moines, 49170 Savennieres, France

TO VISIT BEFORE YOU DIE BECAUSE

This is the site of outstanding Chenin Blanc from a mother-daughter duo.

Madame Monique Laroche created the vineyard in 1981 and over the years the estate's wines have risen to cult-like status. The vines are planted on hillsides with south and southwest exposures, and proximity to the Loire River moderates the normally cold climate. In general, Chenin Blanc is a bit of a chameleon grape and in the schist soils of Savennières it takes on a flinty, chalky, and energetic profile. From the beginning, Laroche worked organically and with her daughter Tessa now leading the way, the family is moving towards biodynamics.

domaine-aux-moines.com

0613225853

86 GRAND CRU LE CLOS – DOMAINE WILLIAM FÈVRE

21 Avenue d'Oberwesel, 89800 Chablis, France

TO VISIT BEFORE YOU DIE BECAUSE

It's one of the first vineyards in Chablis.

The largest of Chablis's seven Grand Cru vineyards at 26 hectares (64 acres), Grand Cru Le Clos is considered the pinnacle vineyard of the appellation. It's also believed to be one of the oldest; evidence proves vines existed in these limestone soils as far back as 1267. The name "Le Clos" refers to the stone wall that once surround the vineyard. While it defined the vineyard's boundaries, it also served more practical purposes: to prevent erosion and to deter thieves—both human and animal—from stealing grapes. Steal a sip for yourself at the William Fèvre wine shop.

williamfevre.fr/en/home +33 (0)3 86 98 98 98

87 L'ERMITE – M. CHAPOUTIER

18 avenue du Dr Paul Durand, 26600 Tain L'Hermitage, France

TO VISIT BEFORE YOU DIE BECAUSE

This tiny vineyard is the pinnacle site for a superstar Rhône producer.

Michel Chapoutier is a big personality, and with wines made in nearly every part of the Rhône, he's a large presence in the wine world. But it's this tiny vineyard that commands the most attention. One-hundred-year-old Marsanne vines and 80-year-old Syrah vines grow in this *lieu-dit* (named site) on the top of Hermitage Hill. The granitic soils are poor in nutrients and vines struggle, yet from this hardship come complex and age-worthy wines (the estate claims up to 75 years). Hospitality plays a major role at the winery; an e-bike tour through vineyards, lunch at the on-site restaurant, and classes at the M. Chapoutier wine school are just a few of the offerings.

chapoutier.com 0760310627

88 LE CLOS LANSON – CHAMPAGNE LANSON

66 rue de Courlancy, 51100 Reims, France

TO VISIT
BEFORE YOU DIE
BECAUSE

This is a secret vineyard in the heart of the city of Reims.

The 1-hectare (about two and a half acre) vineyard Clos Lanson, affectionately known as Lanson's "Secret Garden," sits on a hilltop in the town of Reims, shielded from city life by its surrounding stone wall. With its chalky soils and unique microclimate, the vineyard is the source of one of Lanson's top Champagnes: Clos Lanson. While it's not easy to access the vineyard, Lanson warmly invites visitors to experience a tasting flight in the tasting room or explore the cellars. Other bespoke experiences on offer include vineyard tours and lunch on request.

lanson.com +33 3 26 78 50 50

89 LES THORINS – CHÂTEAU DU MOULIN-À-VENT

Les Thorins, 71570 Romanèche-Thorins, France

TO VISIT BEFORE YOU DIE BECAUSE

A 300-year-old windmill sits among award-winning vineyards.

Rumor has it that King Louis XIV tasted wines from this site and declared them fit for a king. However, it took years of tireless work by Madame Philiberte Pommier, who noticed the full potential of the Gamay vineyards, before Les Thorins truly gained recognition. Her studies in parcel selection, pruning techniques, and aging experiments finally paid off; at the Universal Exhibition in London in 1862, when she was 99 years old, three of her vintages won top awards. Today, the historic stone windmill still stands guard over these highly revered vineyards, while the newly opened tasting room shows the estate's modern side.

chateaudumoulinavent.com

No telephone number

90 MAISON RUINART

4, Rue des Crayères, 51 100 Reims, France

TO VISIT BEFORE YOU DIE BECAUSE

Five miles of underground chalk caves hold both history and wine.

While the magic may start in the vineyard for Ruinart, it is in the underground chalk caves where the wines come to life. Forty meters (131 feet) below ground, the crayères (chalk cellars) run over 8 kilometers (5 miles) under the city of Reims. In the late 18th century, the Ruinart family realized that these cellars—built during the Gallo-Roman period—provided a cool and constant temperature that was ideal for aging wine. Now a UNESCO Heritage site, the maze of cellars provides endless wonderment for visitors. Wander through this underground marvel and hear stories from its history—it served as a bomb shelter during WWI—then emerge aboveground to taste Ruinart's beautifully aged Champagnes.

ruinart.com +33326775152

91 MONTRACHET GRAND CRU – BOUCHARD PÈRE & FILS

2 Rue Jacques Germain, 21200 Beaune

TO VISIT BEFORE YOU DIE BECAUSE

It is considered the home of the best Chardonnay in the world.

When one talks of White Burgundy, the reference is Chardonnay. While this seems simple enough, Burgundy's diverse terroir means Chardonnay expressions vary so widely your head could spin. However, it's the Montrachet Grand Cru in the Côte de Beaune that many point to as the source of the most exquisite Chardonnay. Called *roi de rois* (king of kings) throughout the centuries, the vineyard can be traced back to the 13th century. It is known for its well-draining gravel and limestone soils, into which roots access whatever water remains. Bouchard, one of the largest landowners in the Côte d'Or, is renowned for its Montrachet Grand Cru wines. Taste a legend at their tasting room in Beaune.

bouchard-pereetfils.com/en/home-page +33 (0)3 80 24 80 24

92 ROMANÉE-CONTI – DOMAINE DE LA ROMANÉE-CONTI

1 Pl. de l'Église, 21700 Vosne-Romanée, France

TO VISIT BEFORE YOU DIE BECAUSE

It's the most revered vineyard in the world.

DRC. Three little letters, one big reputation. As it denotes the most coveted and expensive wines in the world, even the name Domaine de la Romanée-Conti sends drinkers into a tizzy. The estate wholly or partially owns eight Grand Cru vineyards, including the crowning jewel of Burgundy: the monopole La Romanée-Conti. A mere 1.81 hectares (4.48 acres), the site produces very little fruit. It could be said that the limestone and clay soils or Burgundy's climate explain why the wines from this vineyard are so exceptional, but that really wouldn't do them justice.

m.romanee-conti.fr

No telephone number

93 SCHOELHAMMER – FAMILLE HUGEL

3 Rue de la Premiere Armee, 68340 Riquewihr, Alsace, France

TO VISIT BEFORE YOU DIE BECAUSE

Here you can witness terroir in action at a vineyard-within-a-vineyard.

Schoenenbourg Grand Cru vineyard already has its own sterling reputation, but within its confines is a small, 30-vine site so spectacular for Riesling that the Hugel family identified it as its own vineyard. Schoelhammer is located in the mid-slope of Schoenenbourg, an ideal spot for any vine. Clay marl soils give a flinty mineral note while the south-facing exposure provides plenty of grape-ripening sunshine. Organic farming ensures the land will continue to thrive for generations to come. Visit Hugel's tasting room to understand why this tiny vineyard produces a wine with such an excellent reputation.

hugel.com +33389479215

94 BRAUNEBERGER JUFFER – WEINGUT FRITZ HAAG

Weingut Fritz Haag, Dusemonder Str. 44, 54472 Brauneberg, Germany

TO VISIT BEFORE YOU DIE BECAUSE

This dramatic slope is home to top Riesling.

For those afraid of heights, turn back now; the almost vertical slope of this Grand Cru vineyard can make you feel as if you are walking off a cliff. But if you make it to the top, it offers some of the most breathtaking views of the Mosel River. The shard-like slate soils retain heat during the day and keep vines warm at night, and also provide an intense minerality to the Riesling. Fritz Haag's Rieslings, especially the Juffer Sonnenuhr, from a small parcel within the greater vineyard, are some of the most intense and complex around. Book a tasting at the winery and try to remain grounded after savoring these stunning wines.

weingut-fritz-haag.de +49 6534-410

95 FORSTER KIRCHENSTÜCK – WEINGUT VON WINNING

Weinstraße 10, D-67146 Deidesheim, Germany

TO VISIT BEFORE YOU DIE BECAUSE

A small stone wall makes a big difference in this vineyard.

Forster Kirchenstück may just be a stone's throw away from a church (*kirche* is German for church), but it is not divine intervention that creates beautiful wines; it's a knee-high stone wall around the vineyard. The structure absorbs heat throughout the day and releases it at night, providing warmth to the vines. In a way, this wall creates a unique microclimate for this 3.67-hectare (9-acre) site that sits adjacent to other Von Winning vineyards. Explore the site on one of the guided vineyard tours, or simply sit back and enjoy a bottle of wine at the Leopold restaurant.

von-winning.de +49 6326 966 870

96 RÜDESHEIM BERG SCHLOSSBERG – WEINGÜTER GEHEIMRAT J. WEGELER

Friedensplatz 9-11, 65375 Oestrich-Winkel, Germany

TO VISIT BEFORE YOU DIE BECAUSE

A defense castle from the 1200s now guards the vines.

The site of Berg Schlossberg was once a perfect lookout spot; it sits at the bend of the Rhine River, allowing multidirectional views. Remains of the Ehrenfels Castle still guard the land, but now it's surrounded by vines, planted on some of the steepest slopes in the Rheingau. In this GG vineyard (the German equivalent of Grand Cru), quartzite and slate soils absorb daytime heat and release it slowly at night, while the river serves as a mirror during daytime hours, ensuring the vines receive ample warmth. Let your guard down and enjoy the vibrant acidity and fruit in the wines at Weingüter Wegeler's tasting room.

 www.wegeler.com +49 6723 99090

97 ÜRZIGER WÜRZGARTEN – DR. LOOSEN ESTATE

St. Johannishof 1, D-54470 Bernkastel/Mosel, Germany

TO VISIT
BEFORE YOU DIE
BECAUSE

This is an amazing Riesling "spice garden."

You can't miss Ürziger Würzgarten: its blazing red, iron-rich soils stand out from other vineyards of the Middle Mosel. Due to the high iron content, Riesling from this site takes on herbal and spice aromas that are very distinct from other Mosel Rieslings. Most of the old vines are ungrafted, meaning they continue to grow on original rootstock. At Dr. Loosen's family-run estate with its 200 years of history, see if you can identify this spice garden wine in a tasting flight, or book a weekend stay at the guesthouse.

drloosen.com

+49 (0) 6531-3426

98 ALPHA ESTATE VINEYARDS – ALPHA ESTATE

2nd Klm. Amyndeon, Ag. Panteleimon. 53200 Amyndeon, Greece

TO VISIT BEFORE YOU DIE BECAUSE

Snow falls on this Greek vineyard in winter.

Although most of Greece's wine appellations are considered to have a Mediterranean climate (warm, sunny, and dry growing seasons with moderate winters), the area of Amyndeon stands out for its cold winters and snowfall. This semi-continental climate at Alpha Estate's 180-hectare (445-acre) vineyard contributes to the award-winning wines. The water provided in the colder months nourishes the vines throughout most of the dry summer, allowing for the development of both indigenous and international grape varieties, such as Xinomavro, Mavrodapnhe, Syrah, Merlot, Pinot Noir, Montepulciano, Malagouzia, Sauvignon Blanc, and Chardonnay. No matter whether you prefer snow or sun, you won't regret a tour of the vineyards.

 alpha-estate.com/ +30 23860 20111

99 ESTATE VINEYARDS – KTIMA KIR-YIANNI

Yianakohori, GR–59200 Naoussa, Greece

TO VISIT BEFORE YOU DIE BECAUSE

The estate is the personal project of a former wine company executive.

Yiannis Boutaris was no stranger to wine; his family owned one of the largest Greek wine companies, Boutari Wine Group, established in 1879. But Yiannis wanted more, so he set out to create a boutique winery in Naoussa, focusing on indigenous varieties, particularly Xinomavro and international grapes such as Syrah and Merlot. Today, Yiannis's son, Stelios, oversees production and continues to push the boundaries of Xinomavro's potential. Explore the family's passion through picnics, bike rides, or an open-air movie night among the vines.

kiryianni.gr +30 23320 51100

100 PYRGOS VINEYARD – GAI'A WINES

Nemea Peloponnese, Santorini, Cyclades, Greece

TO VISIT BEFORE YOU DIE BECAUSE

Life is not a beach on Santorini's rugged terrain.

The volcanoes of Santorini not only created the island's stunning black sand beaches but also a unique—if unlikely—terroir for wines. In the Pyrgos vineyard, the soils bring a strong mineral character to GAI'A's wines. Because there is little groundwater on the island itself, vineyards are planted at an extremely low density to ensure each vine gets enough to thrive. Strong winds regularly batter the island, and vines are trained to weave together in a low-to-the-ground basket shape. Enjoy the fruits of all this labor at the GAI'A stunning seasonal tasting room in Santorini with its views of the Aegean Sea.

 gaiawines.gr 00302108055642

101 PETRACS VINEYARD – TOKAJ-OREMUS

Bajcsy-Zsilinszky út 47, Tolcsva 3934, Hungary

TO VISIT BEFORE YOU DIE BECAUSE

This vineyard was formerly owned by a Major General of the Imperial Army of the Austro-Hungarian Empire.

Baron Ernest von Petrasch was a leader in the Austro-Hungarian army, and his legacy lives on in the name of his vineyard, which he acquired in 1762. The wines that come from this site are equally legendary: the bedrock and stony soils give wine a great minerality, while the 60-year-old vines provide depth and concentration. Because of the steepness of the vineyard, all grapes are handpicked as only humans and horses can navigate the rows.

temposvegasicilia.com +36 47 384 520

102 MÉZES MÁLY – ROYAL TOKAJI

Rákóczi u. 35, H-3909 Mád, Hungary

TO VISIT BEFORE YOU DIE BECAUSE

This is the source of one of the greatest sweet wines in the world.

One of the only two vineyards to be named a "Great First Growth" in the 1700 classification, Mézes Mály (loosely translating as "honeycomb") is prized for the honey and floral character of its wines. While soil plays a part in the wine's character—the loess topsoil gives a soft texture—there's another factor at play, namely rot. In order to make sweet wine, the Furmint, Hárslevelü, and Muscat de Lunel grapes need to be infected by botrytis, also known as "noble rot." These infected grapes, called *aszú*, go through a rigorous selection and fermentation process. Because of Mézes Mály's unique character, it is bottled as a single-vineyard sweet wine. Sweeten your wine trip with a tasting appointment at the estate.

royal-tokaji.com +36 47 548 500

103 ALOIS LAGEDER

Vicolo dei Conti 9, 39040 Magrè, Italy

TO VISIT
BEFORE YOU DIE
BECAUSE

This is where wild oxen act as natural lawnmowers.

The Alois Lageder estate has been a leader in biodynamics for decades, and one of their most charming practices is their oxen project. Every autumn, cowbells ring out as cattle are brought down from the Käserei Englhorn organic cheese dairy in Val Venosta to the estate's vineyards for the winter. The warmer climate in southern Alto Adige ensures there is food for the cattle to graze on all season long. Not only does it solve the problem of food scarcity in the Val Venosta, but the cows' presence creates greater biodiversity in the vineyard; their manure provides extra nutrients to the soil. Taste the fruits of the cows' labor at Paradeis, the all-organic wine bar and restaurant run by the Lageders.

aloislageder.eu +39 0471 809 500

104 COLONIA – FÈLSINA

4 Via Del Chianti 101, Castelnuovo Berardenga, Siena, Italy

TO VISIT BEFORE YOU DIE BECAUSE

It took three generations to bring this vineyard to fruition.

Starting in 1966, Domenico Poggiali and his son Giuseppe struggled to create a vineyard in the rocky hilltop terrain of Rancia. They persevered for a couple of decades but ultimately admitted defeat. In 1991, Giovanni picked up the torch and spurred his father and grandfather to have another go at the land. In 1993 they completed the project. Although Domenico didn't live long enough to see the first vintage, this 100 percent Sangiovese wine is considered the estate's top bottling—a fitting tribute to the generations behind its genesis. Luckily, tasting the wine is much easier than making it, so book a tour and a tasting at the estate's wine shop.

felsina.it

+39 0577355117

105 ESTATE VINEYARDS – ORNELLAIA

Località Ornellaia, 191, Fraz. Bolgheri, 57022 Castagneto Carducci (LI), Italy

TO VISIT BEFORE YOU DIE BECAUSE

Bordeaux varieties take on a whole new expression in this Italian terroir.

The coastal appellation of Bolgheri is an anomaly in the Sangiovese-dominant region of Tuscany. Here, Bordeaux varieties such as Cabernet Sauvignon and Merlot thrive, thanks to the ocean's moderating effects on the climate. One of the leaders of this style of wine is Ornellaia, founded in 1981. Because each vintage is considered a work of art, the estate established its Vendemmia d'Artista project in 2009. Each year a different artist creates an artwork which is featured on limited-edition labels. See creativity on display as you tour the winery's on-site gallery.

ornellaia.com +39 0565 71811

106 ESTATE VINEYARDS – ABBAZIA DI NOVACELLA

Via Abbazia 1, 39040 Varna (BZ), Alto Adige, Italy

TO VISIT BEFORE YOU DIE BECAUSE

A working monastery is the source of great wines.

Perhaps it's the presence of the church on the property, but it's apparent that Abbazia di Novacella is no ordinary winery. In fact, it's part of a working monastery which originated in 1142. However, winemaking has always been part of its history almost since it was founded; Abbazia di Novacella has owned vineyards since 1177 and is one of the oldest working wineries in the world today. Estate vineyards, planted with white varieties such as Sylvaner, Müller Thurgau, Kerner, Grüner Veltliner, Pinot Grigio, Riesling, and Gewürztraminer surround the property. Spend a day at the estate touring the museum and visiting the winery, then enjoy a tasting and lunch at the restaurant. (I recommend the apple strudel—it's delicious.)

abbazianovacella.it +39 0472 836189

107 ETNA ESTATE – DONNAFUGATA

Contrada Statella, 95036 Randazzo (CT), Italy

TO VISIT BEFORE YOU DIE BECAUSE

Volcanic wines from the slopes of Mount Etna are explosively good.

An active volcano may seem an unusual place to plant a vineyard, but the volcanic soils and high elevation of Mount Etna create ideal growing conditions for indigenous Sicilian grape varieties. Small bush vines of 70-year-old Carricante, Nerello Mascalese, and Nerello Cappuccio hug the ground while walls and terraces of lava stone (awarded UNESCO Intangible Cultural Heritage status) create order in the vineyard. Be prepared to be blown away by the beauty of the vineyards during a walking tour before cooling down with sips of wine at the wine shop.

donnnafugata.it

+39 095 7991949

108 FAEDO – POJER E SANDRI AZIENDA AGRICOLA

Via Molini 4, 38098 San Michele all'Adige, Faedo (TN), Italy

TO VISIT
BEFORE YOU DIE
BECAUSE

This ancient site inspired some of the most inventive wines in Trentino.

The almost conical shape of Faedo in Trentino is credited to glacial landslides, and the river Faedo further helped the downward trajectory of sediment. Along the way, numerous soil types settled. Combined with long hours of sunshine and cool breezes, Pojer e Sandri saw it as the perfect place to experiment. They cultivate the rare variety Nosiola and elevate Müller-Thurgau (which can be rather insipid) to a fresh and energetic white wine. In the cellar they produce two styles of sparkling wine along with the lineup of white wines, and continue to try out and incorporate new vinification processes, such as a grape "bath." Put on your hiking shoes for a tour of the vineyard, then head back to your room at their B&B for a little rest and relaxation.

pojeresandri.it

+39 0461 650342

109 FONDAZIONE WHITAKER – TASCA D'ALMERITA – TENUTA WHITAKER

Isola di Mozia, Marsala (TP), Italy

TO VISIT BEFORE YOU DIE BECAUSE

The uninhabited island of vines is home to well-preserved Phoenician archeological sites.

Phoenicians settled on the island of Mozia in the 8th century BC once they realized that the shallow Marsala lagoon which surrounded the island provided ample protection, while allowing inhabitants to easily flee from invaders. Soon, a rich culture developed, which included winemaking. In 2007, after centuries of abandonment, a new project aimed to redevelop the ancient land and plant vineyards of the native variety Grillo. During the revitalization, numerous Phoenician artifacts were recovered. Sail to the island, visit the small Phoenician museum, and view this new, vinous landscape.

tascadalmerita.it

+390916459711

110 GORGONA – FRESCOBALDI

Ristorante Frescobaldi Firenze, Piazza della Signoria 31, 50122 Firenze FI, Italy

TO VISIT BEFORE YOU DIE BECAUSE

The vineyard is cultivated on a prison island.

With its sweeping ocean views and balmy Mediterranean climate, it's hard to believe this paradise-like island houses a prison. But Gorgona, an hour-and-a-half boat ride from the Tuscan coastal town of Liverno, is indeed part of the Italian prison system. A unique social project in partnership with the Frescobaldi family provides inmates with valuable skills in grape growing and a potential career in the wine industry once they are released. From the 3.5-hectare (9-acre) vineyard of Ansonica and Vermentino grapes, prisoners learn viticulture and viniculture, and earn a wage so they can re-enter the world with savings. Given the size of the vineyard, only 9,000 bottles are produced every vintage, but you can taste this delicious white wine for yourself at Frescobaldi restaurant in Florence.

frescobaldifirenze.it

No telephone number

111 IL GREPPO – BIONDI-SANTI

Villa Greppo 183, 53024 Montalcino (Siena), Italy

TO VISIT BEFORE YOU DIE BECAUSE

The birth of a wine region originated in this vineyard.

Brunello di Montalcino DOCG can trace its entire existence back to the Il Greppo vineyard and the work of several generations of the Biondi-Santi family. In 1865, Clemente Santi won an award for the first-ever Brunello-labeled wine at the 1867 Universal Exposition in Paris. His experiments with aging Brunello wine provided direction for the entire region. Decades later, under third-generation Ferruccio Bondi, the Sangiovese clone BBS 11 (Brunello Biondi-Santi) was propagated at Il Greppo. Bondi also made the decision to bottle only 100 percent Sangiovese wine, which is now a regulation for Brunello DOCG wines. For those lucky enough to own Riserva, the estate recently reintroduced its Ricolmatura Il Greppo program, where it tops up old bottles in which a bit of wine has evaporated and seals them with new corks.

biondisanti.it/ +39 0577 848023

112 ORNATO ESTATE VINEYARD – PIO CESARE

Via cesare Balbo 4, 12051 Alba, Italy

TO VISIT BEFORE YOU DIE BECAUSE

Here you can taste the very first single-vineyard bottle from a Piedmont legend.

Less than 5 percent of the Piedmont region is flat, meaning the patchwork of vineyards faces multiple directions and sits at multiple altitudes. This diversity comes together at Ornato, the site of Pio Cesare's first single vineyard Barolo. The 6.6-hectare (16-acre) amphitheater-shaped vineyard captures the sun throughout the day with its south, southeast, and southwest exposures. From the predominately limestone and clay soils, Nebbiolo develops a lot of structure and tannins, resulting in a wine that's meant to age for years. Dig deep into history during a tour of the winery's ancient cellars, which are surrounded by 2,000-year-old Roman walls in the town of Alba.

piocesare.it

+390173440386

113 MARTINENGA VINEYARD – TENUTE CISA ASINARI DEI MARCHESI DI GRÉSY

Località Martinenga, Strada della Stazione 21,
12050 Barbaresco (CN), Italy

TO VISIT BEFORE YOU DIE BECAUSE

This rare monopole vineyard is bottled by a single producer.

The amphitheater-shaped vineyard, property of the Marchesi de Gresy family since 1797, is a true study in terroir. Designated a Menzione Geografica Aggiuntiva (MGA), it is the equivalent of a Grand Cru vineyard elsewhere in the world. Although one might expect a single vineyard to create a singular style of wine, the various exposures and altitudes of the vineyard allow Marchesi di Gresy to create three distinct wines to capture the different expressions of Martinenga: Gaiun, from the western vines; Martinenga, from the middle vines; and Camp Gros from the eastern plots. Book a tasting and try all three for yourself, or stay at Dai Gresy, the newly-opened hotel and spa and spend a little longer pondering the versatility of Martinenga.

marchesidigresy.com +30 0173635222

114 RUFFOLI ESTATE VINEYARDS – QUERCIABELLA

Via di Barbiano 17, 50022 Greve in Chianti (FI), Italy

TO VISIT BEFORE YOU DIE BECAUSE

From a single hectare on a Tuscan hillside came one of Italy's leading biodynamic and vegan wine producers.

In 1974, entrepreneur Giuseppe Castiglioni made a life change with the purchase of 1 hectare (two and a half acre) of vines in the Tuscan village of Ruffoli. His dream of growing Sangiovese and Bordeaux grapes, such as Cabernet Sauvignon, came to fruition, but it was his son, Sebastiano, who elevated Querciabella's standing. In 1988 he converted the vineyard to organic farming and in the early 2000s became biodynamic. He also banned all animal products from the vineyards and cellars, establishing Querciabella as one of the first publicly recognized vegan wineries.

querciabella.com

+39 055 8592 7777

115 SORI TILDÌN – GAJA

Via Torino 18, 12050 Barbaresco, CN, Italy

TO VISIT BEFORE YOU DIE BECAUSE

This vineyard is a homage to an influential family figure.

Angelo Gaja is recognized as a pioneer in the Piedmont region—he was one of the first to use 225-liter French barrique barrels, bottle single-vineyard wines, and plant international grapes such as Cabernet Sauvignon and Chardonnay—and he credits his inspiration to his grandmother, Clotilde Ray. In honor of her influence, this 3.2-hectare (8-acre) site carries her nickname, Tildìn. But there's nothing diminutive about this site, considered one of the greatest vineyards in all of Barbaresco. Although it is not technically open to the public, guests who make a charitable donation to one of Gaja's selected nonprofit organizations will be given a two-hour tour of the cellar and a tasting in appreciation of their contribution.

gaja.com

+39 0173 635158

116 VENISSA TENUTA VINEYARDS – VENISSA

Fondamenta Santa Caterina 3, 30142 Mazzorbo, Venezia, Italy

TO VISIT BEFORE YOU DIE BECAUSE

You can visit an island vineyard in the middle of a Venetian lagoon.

Vineyards existed on the islands in the Venetian lagoons for centuries due to their ability to adapt to saltwater and high tides. However, in the flood of 1966, nearly all viticulture was wiped out, save for 88 vines. When Gianluca Bisol found these vino vestiges, he set out to revive the wine-making tradition of the area. He established what was to become Venissa on the island of Mazzorbo and planted the estate with Dorona di Venezia, a variety known for withstanding the particular conditions of the lagoon. Enjoy Venissa's work with vineyard tours, wine tastings, and culinary excursions to the estate's two restaurants.

venissa.it

+39 041 5272281

117 VIGNA CASTELAZ – ELENA WALCH

Via Andreas Hofer 1, Tramin, Alto Adige, Italy

TO VISIT BEFORE YOU DIE BECAUSE

This vineyard, farmed by a mother-and-daughters team, produces outstanding Gewürztraminer and Merlot.

When architect Elena Walch married into a winemaking family, she didn't know she would end up redesigning her career path. She modernized the estate and her now-eponymous winery, run by her and her daughters Karoline and Julia, is one of the most respected in Alto Adige. Vigna Castelaz, with its steep slopes, south-facing exposure (unusual for the region), and layered soils, is a prime spot for the white variety Gewürztraminer and red variety Merlot. Walch's design background can be seen at the Garden Bistrot, where you can taste the wines and enjoy a few bites after a vineyard tour.

elenawalch.com 00390471860172

118 ESTATE VINEYARD – DE KLEINE SCHORRE

Zuiddijk 4, 4315PA Dreischor, The Netherlands

TO VISIT BEFORE YOU DIE BECAUSE

You can sleep in a wine barrel house overlooking the vineyards.

At De Kleine Schorre in the Netherlands, wine is a way of life. The varieties the estate cultivates in the 14-hectare (35-acre) vineyard—Pinot Gris, Pinot Blanc, Rivaner, and Auxerrois—were chosen specifically to pair with the fish, shellfish, and other salty fare common to the region. To really integrate into the wine experience, book a stay at one of the wine barrel houses where you can wake up in the morning and be greeted by the vineyards surrounding you.

dekleineschorre.nl 0111 – 40 15 50

119 ESTATE VINEYARDS – FILIPA PATO AND WILLIAM WOUTERS

Rua de Santo André, 41, 3780-502 Óis do Bairro
Anadia, Portugal

TO VISIT BEFORE YOU DIE BECAUSE

A second-generation winemaker and a restaurateur teamed up to create cutting-edge Portuguese wine.

As the daughter of rebel winemaker Luis Pato (champion of old-vine Baga), winemaking is in Filipa Pato's DNA and she applied her degree in chemical engineering to her various winemaking stints throughout the world. Her husband, William Wouters, brings his expert palate (he's a restaurateur and sommelier) to their project. Together they highlight the indigenous varieties of the Bairrada region—Baga, Bical, Arinto, Cercial, and Maria Gomes—while working in a low-intervention method that creates, as Filipa puts it, "wines without makeup." Make an appointment for a tasting of these refreshing renditions of native grapes.

patowouters.com +35 1231516041

120 ESTATE VINEYARDS – QUINTA DA AVELEDA

Rua da Aveleda, Nº2, 4560-570 Penafiel, Portugal

TO VISIT
BEFORE YOU DIE
BECAUSE

You can wander among the lush gardens of this historic estate.

Quinta da Aveleda is a botanical wonderland with lush, leaf-covered vines and a beautiful garden with wandering paths and numerous specimens of flora and fauna. From the Japanese cedar trees to blooming camellias, the garden bursts with color and life. Take a moment of quiet contemplation at the tea house and pavilion that sits on the lake, or book a guided tour and tasting to ensure you don't miss any of the treasures hiding in this not-so-secret garden.

aveleda.com +35 1255718242

121 HERDADE DE ESPORÃO

Herdade do Esporão, Apartado 31, 7200-999 Reguengo de Monsaraz, Portugal

TO VISIT BEFORE YOU DIE BECAUSE

This property is home to medieval architectural marvels.

At the site of Herdade de Esporão and dating back almost 600 years, the historical Esporão Tower, the 16th century Nossa Senhora dos Remédios Chapel, and the Esporão Arch stand as reminders of the land's history. The structures aren't the only things being preserved; Esporão farms the 700 hectares (1,730 acres) of vineyards organically to ensure their longevity. There's much to discover at the estate. Tour the historical monuments, take a bike ride through the vineyards, or dine at the on-site Michelin-starred restaurant.

esporao.com (+351) 266 509 280

122 QUINTA DA LIXA – MONVERDE WINE EXPERIENCE HOTEL

Quinta de Sanguinhedo 166, Castanheiro Redondo, 4600-761 Telões, Amarante, Portugal

TO VISIT BEFORE YOU DIE BECAUSE

You can get a massage in the middle of a vineyard.

Quinta da Lixa got its start selling wines at its cellar doors to the locals of Vila da Lixa. As word spread, winemaking became a full-blown affair for the Meireles family and they quickly acquired estate vineyards to support production. In 2015, they opened Monverde Wine Experience Hotel nearby. Set in the middle of vineyards, it's an ideal place to immerse yourself in wine-related experiences. Book a massage at the spa and take in the natural bounty of the vineyards that surround you.

quintadalixa.pt

No Telephone number

123 QUINTA DE RORIZ – PRATS & SYMINGTON

Quinta de Roriz, S. J. da Pesqueira, 5130-113 Ervedosa do Douro, Portugal

TO VISIT BEFORE YOU DIE BECAUSE

Tailings from old mines influence the terroir of this dramatic site.

Quinta de Roriz rises dramatically from the banks of the Douro River and the amphitheater-shaped vineyard creates a breathtaking backdrop as you pass by via boat. A special addition to the terroir gives these wines extra shine: remnants from old gold and tin mines above the estate imbue them with a particular minerality and flavor. Travel the Douro River to another estate of the Symington family, Quinta do Bomfim, where guided vineyard walks, tastings, and tours are on offer at the visitor center and restaurant.

chryseia.com

+351 223776300

124 SOALHEIRO

Lugar da Charneca, 4960-010 Alvaredo, Melgaço, Portugal

TO VISIT
BEFORE YOU DIE
BECAUSE

The vineyard started an Alvarinho renaissance in Monção and Melgaço.

Although grapes have been cultivated in Monção and Melgaço for centuries, winemaking was not considered an important crop and premium land was used for other agricultural endeavors. In 1974, the Cerdeira family departed from the traditional way of thinking and planted a full vineyard of Alvarinho in the granitic soils. Their intuition paid off and the white variety thrived in this parcel, named Soalheiro, meaning "sunny place" in Portuguese. Others in Monção and Melgaço took their cues from the family's work and now high-quality Alvarinho is a signature of the area. Always forward-thinking, Soalheiro continues to experiment with the grape's potential, from amphora aging to sparkling *pétillants naturels*. See where it all began during a vineyard tour of the property.

soalheiro.com +351 251416769

125 QUINTA DE VARGELLAS – TAYLOR FLADGATE

Rua do Choupelo, nº 250 4400-088 Vila Nova de Gaia, Portugal

TO VISIT
BEFORE YOU DIE
BECAUSE

You can stomp grapes in a traditional lagare.

The hilly, terraced vineyards are the source of Taylor's finest single-estate ports, including a tiny production of Vinha Velha vintage port, made from five separate plots of old vines on the property, some of which are over 100 years old. The winery continues to respect the region's past and traditions; after harvest, grapes are put into large stone tubs called lagares, and teams of people work tirelessly to stomp the grapes by foot. This orderly stepping turns into a full-on dance party as night descends. Wade deeper into your port-wine education at Taylor's Visitor Center in the city of Vila Nova de Gaia. Also on offer are tastings, masterclasses, and pairing meals at the on-site restaurant.

126 FOJANA VINEYARD – EDI SIMČIČ WINE ESTATE

Vipolže 39a, 5212 Dobrovo, Slovenia

TO VISIT BEFORE YOU DIE BECAUSE

This vineyard is the source of exceptional Rebula.

Rebula (also known as Ribolla Gialla in Italy) is a grape for wine lovers in the know; it's incredibly versatile, while also capable of producing age-worthy and structured white wines. Much of its production is centered around Brda in Slovenia and Collio in Italy, and one of the best examples comes from the Fojana Vineyard in Slovenia. The hillside vineyard contains rocky and opoka soils, (layered sandstone and marl) which give a savory minerality to the wine. The sunny exposure allows grapes to fully ripen and complex fruit flavors to develop. At Edi Simčič Wine Estate taste Fojana's influence on this white variety during a tasting and winery tour.

edisimcic.si/vina_en +38630602564

127 AMEZTOI – BODEGA DE TXAKOLI AMEZTOI

Barrio Eitzaga, 10 - Apdo. 246, 20808 Getaria (Gipuzkoa), Spain

TO VISIT BEFORE YOU DIE BECAUSE

This is where Txakoli Rosé was first created.

Txakoli—a light, crisp, and slightly frizzante white wine—has been a staple in the Basque region for years, but in 2005 a new wine made headlines: Txakoli Rosé. Behind this creation was Ignacio Ameztoi, whose family had been making wine in their small fishing village for five generations. Txakoli is made from the white grape hondarrabi zuri, but the family owned a small plot of 150-year-old hondarabbi beltza vines (a red grape). Ameztoi came up with the idea of blending the two, and the cult rosé wine was born. Book a stay at the boutique hotel or take a tour of the winery, and life is sure to be rosy.

txakoliameztoi.com 943 140918

128 CONTINO ESTATE VINEYARDS – VIÑEDOS DEL CONTINO

Finca San Rafael (LaSerna), 01300 Laguardia, Alava, Spain

TO VISIT BEFORE YOU DIE BECAUSE

This is the site of Rioja's first single-vineyard wine.

For centuries, Rioja wines relied on the blending of fruit from multiple vineyards for their wines, but in 1973, family-owned CVNE had the vision to highlight one particular site: Contino. This south-facing site, with its chalk, sand, and clay soils, was a key source of grapes for CVNE's Viña Real Gran Reserva. However, it became apparent that grapes from the site deserved to be a wine of their own, and thus Contino was born. Over time, the vineyard was devided into 32 micro-parcels, allowing for even more experimentation. Contino now makes a single-varietal Graciano (which is normally used as a blending grape) and a white wine made from Viura, among others. Tour this treasure trove of a vineyard, then taste its offerings at the wine bar.

cvne.com/en/wineries/vinedos-del-contino/ +34 945 600 201

129 FINCA BERMEJO – BODEGA BERMEJO

Camino a Los Bermejos, 7, 35550 San Bartolomé, Spain

TO VISIT BEFORE YOU DIE BECAUSE

Black volcanic soils feel otherworldly.

Looking out at the blanket of black ash soils, it seems incomprehensible that grapes grow well here. But a visit to Finca Bermejo in the area of La Geria on the island of Lanzarote reveals the site of delicious white and red wines. Vines grow in deep holes to protect them from the strong winds. These moon-like craters are as deep as 4 meters (13 feet) and as wide as 6 meters (almost 20 feet) in diameter. Because the deadly vine disease phylloxera cannot thrive here, many of the old vines are planted on their original rootstock. A visit to Bodega Bermejo yields even more surprises: a guided tour of the vineyards could end with a spontaneous blind tasting of the estate's savory and saline-tinged wines, such as Malvasia and Listan Negro.

losbermejos.com +34 928 522 463

130 ESTATE VINEYARDS – CLOS MOGADOR

Clos Mogador, SL Camí Manyetes, s/n, 43737-Gratallops, Spain

TO VISIT BEFORE YOU DIE BECAUSE

Here you can visit one of the original "Gang of Five" that put Priorat on the map.

Like many wine regions that were destroyed by phylloxera, Priorat's recovery was in doubt for many years. But in 1979, Rene Barbier Ferrer purchased a vineyard and others soon followed. A group of entrepreneurs, nicknamed the "Gang of Five," named their sites "Clos" in the same manner as the French, a revolutionary idea at the time. Garnacha and Carignan became the grapes du jour in these incredibly steep vineyards, and soon the quality elevated Priorat to the upper echelon of the world's wine regions. Tour the clos where it all began and taste the evolution of the region with a flight of Clos Mogador's wines during a visit to the winery.

closmogador.com +34 977 839171

131 LA CRUSES VINEYARD – THE HOUSE OF LUSTAU

Calle Arcos 53, Jerez de la Frontera, Cádiz, Spain

TO VISIT BEFORE YOU DIE BECAUSE

This is one of the few vineyards in Andalucia where three major soil types exist.

To understand the wines of Jerez-Xérès-Sherry, one needs to understand the soils in which Palomino, Pedro Ximénez, and Moscatel grow. *Albariza,* a chalky-white soil, rich in limestone, is the dominating force which gives the neutral Palomino grape its character. Meanwhile, clay-rich *barros* soils can be found in low-lying valleys, while sandy *arenas* soils, best suited for Moscatel, are found in coastal areas. The flat expanse of La Cruses, located just 3 kilometers (1.86 miles) from the ocean, is one of the rare places that contain all three soils. Watch the ground transform beneath you on a VIP tour, then taste the exceptional sherries from the vineyard.

lustau.es +34 956 34 15 97

132 MAS DE LA ROSA – VALL LLACH

Celler Vall Llach, Carrer Pont 9, 43739 Porrera (Tarragona)
Catalunya, Spain

TO VISIT
BEFORE YOU DIE
BECAUSE

You need to sit and slide in order to harvest this treacherous site.

The name Rose's Farmhouse may sound romantic, but farming the site is not for the faint of heart. The source of Vall Llach's famous single vineyard wine is pitched at a steep 80-degree angle and covered in slate rock. The old vines produce very little fruit; it takes seven vines to get enough grapes for just one bottle. Harvest is done by hand, and pickers often need to sit on their bottoms and slowly scoot down the terrain to obtain the grapes. The effort is worth it; take a seat in the tasting room and enjoy a sip.

valllach.com +34660174308

133 LA RIOJA ALTA, S.A.

Avenida Vizcaya 8, 26200 Haro, La Rioja, Spain

TO VISIT BEFORE YOU DIE BECAUSE

This is the original property of a Tempranillo specialist.

La Rioja Alta, S.A. is known throughout the world for its exceptional Tempranillos, and it all started in 1890 with five families and one wine. Over the years, the company established three other wineries throughout Spain, but it's the original in Haro's Station District that's home to some of the most exquisite Gran Reserva Tempranillo wines in the world. Produced from 60-year-old vines grown on estate vineyards, these age-worthy reds continue to garner acclaim. Book a tour at the venerable estate, then grab a drink at the terrace wine bar.

riojalta.com +34941310346

134 OTAÑON VINEYARD – BODEGAS VEGA SICILIA

Finca Vega Sicilia, 47359 Valbuena del Duero, Valladolid, Spain

TO VISIT BEFORE YOU DIE BECAUSE

Pre-phylloxera vines and unknown grapes contribute to one of the most celebrated Spanish wines.

While many point to Vega Sicilia's long and complex aging methods as a defining characteristic of their renowned Unico, it's really the vineyard that plays the biggest role in the singular character of the wine. At Otañon, old vines that predate phylloxera continue to produce complex and concentrated fruit. However, the vineyard also holds a secret: unidentified grape varieties—possibly mutations or clones of Tempranillo, possibly grapes that became extinct elsewhere—thrive in the soils and are believed to sometimes make their way into the wine.

temposvegasicilia.com +34 928 522 463

135 TIO PEPE VINEYARD – BODEGAS TIO PEPE

Calle Manuel Maria Gonzalez, 12, Jerez de la Frontera 11403, Spain

TO VISIT BEFORE YOU DIE BECAUSE

This is a true sherry legacy.

Sherry is one of the most complex wines in the world to produce: it spends years traveling through a barrel-aging system called a solera, where different lots and vintages are continually blended together. Bodegas Tio Pepe is one of the most recognizable names when it comes to these floral, salty, and nutty wines from the south of Spain. Numerous opportunities abound to experience the wines; book a stay at the Sherry hotel, tour the bodega and its pyramids of barrels, or take in a flamenco concert at the event space.

tiopepe.com

+34 956 35 70 16

136 ESTATE VINEYARD – ARILDS VINGÅRD

Lillavägen 71, 263 73 Arild, Sweden

TO VISIT
BEFORE YOU DIE
BECAUSE

A single vine and a passion turned into one of Sweden's most exciting wineries.

When the Ivarsson family moved to their farm in Kullaberg, a single vine on the property got them thinking about the potential of a new project: a vineyard. After studying wine and traveling to Europe to visit vineyards, owners Annette and Jonas planted their first vines in 2008. Today, they cultivate 20 hectares (50 acres) of Solaris, Pinot Noir Précoce, Cabernet Cortis, Souvignier Gris, and Muscaris. Their home is now a getaway for Scandinavian wine lovers. Go glamping in the vineyards or book a room at the wine hotel.

arildsvingard.se/home 042346420

137 LES BERNUNES – CAVE CALOZ

Ancienne Route de Sierre 1, 3972 Miège, Switzerland

TO VISIT BEFORE YOU DIE BECAUSE

Here you can taste wine from the first vineyard in the Valais region.

Wine lovers don't often get the chance to drink Swiss wines because 98 percent of production is consumed domestically. But for those that visit the tiny wine-producing country, one label to seek out is Cave Caloz Sandrine, Anne-Carole & Conrad. Les Bernunes is the first vineyard ever mentioned in ancient records of the Valais region. The vineyard is situated on high slopes with its own microclimate, and the Savagnin-based wines from the site are truly memorable. Get your fill of Cave Caloz and other Swiss wines at annual events, such as the Open Cellars of Valais Wines.

cavecaloz.ch

No Telephone number

138 VINKARA – VINKARA WINERY

Halit Cevri Aslangil, 06870 Kalecik/Ankara, Turkey

TO VISIT BEFORE YOU DIE BECAUSE

Native grapes are grown in a land with 7,000 years of winemaking history.

There is evidence that wine has been made in Anatolia (modern-day Turkey) for nearly 7,000 years, and anywhere from 600 to 800 indigenous varieties still thrive today. Vinkara's mission is to continue the cultivation of these native grapes in the 81 hectares (200 acres) of vineyards. Four varieties are the focus: the red grapes Kalecik Karasi, Öküzgözü, and Boğazkere, and the white grape Narince. Sit among history and enjoy a picnic in the vineyards.

vinkarawines.com 2038959869

139 APPLEJACK VINEYARD – GIANT STEPS

314 Maroondah Hwy, Healesville VIC 3777, Australia

TO VISIT
BEFORE YOU DIE
BECAUSE

A temperate rainforest aids in the creation of standout single-vineyard Pinot Noir.

The winemakers at Giant Steps say Applejack Vineyard was destined to create single-vineyard Pinot Noirs; with five different clones, several root-stocks, and various spacings of vines, there are numerous expressions of Pinot for them to play with. But none of these vines would thrive if it were not for their proximity to dense temperate rainforest. Cold air descends on Applejack Vineyard at night and cools down the grapes from the heat of the day. Explore the rainforest effect at the winery's cellar door.

giantstepswine.com.au/ +61 3 5962 6111

140 D'ARENBERG

58 Osborn Rd., McLaren Vale, SA 5171, Australia

TO VISIT BEFORE YOU DIE BECAUSE

The multi-sensory experiences at The Cube take wine appreciation to a whole new level.

You see it from a distance, but you aren't sure what you are looking at. Yet you are drawn in. That's the magic of the d'Arenberg Cube, a five-story winery/tasting room/art gallery/dining venue in the middle of the vineyards. Wander through the maze of floors and rooms that combine art and wine education. As one of Australia's most renowned wineries, they've certainly got the know-how. Take one of the wine masterclasses, or play winemaker for a day, complete with a blending and bottling session of your own wine. When you get hungry, book a table at one of several restaurants on the property.

darenberg.com.au

+61 8 8329 4888

141 HILL OF GRACE VINEYARD – HENSCHKE

Henschke Cellars, 1428 Keyneton Rd., Keyneton SA 5353, Australia

TO VISIT BEFORE YOU DIE BECAUSE

These 160-year-old vines are a living legend.

Imported from Europe and planted in the 1860s, these gnarled yet magnificent pre-phylloxera vines produce some of Australia's most beguiling Shiraz. The site is only 4 hectares (10 acres), and coupled with the age of the vines, very little fruit is produced. In the past few decades, the Henschke family has moved towards organic and biodynamic viticulture to ensure the longevity of the vineyard, which is considered the soul of the estate. You can grace the vineyard yourself by booking a vineyard tour that includes a private tasting.

henschke.com.au

+61 8 8564 8223

142 HOME VINEYARD – VASSE FELIX

Corner of Tom Cullity Drive and Caves Rd., Cowaramup, Western Australia

TO VISIT BEFORE YOU DIE BECAUSE

This is the site of Margaret River's first vines.

After reading research papers advising on the development of vineyards in Margaret River, Dr Thomas Cullity planted the region's first vines in what is now called the Home Vineyard. The first vintage in 1971 was a disaster, but a small production of Riesling in 1972 won a gold medal at a show in Perth, and wine lovers took notice. More awards and acclaim followed, and the Margaret River wine region flourished. Today, the Home Vineyard is focused on growing Cabernet Sauvignon and Chardonnay. At the epicenter of the Vasse Felix estate, make yourself at home by enjoying a meal at the award-winning restaurant, exploring the art gallery, or touring the vineyards and cellar.

vassefelix.com.au +61 8 9756 5000

143 MCINTYRE VINEYARD – MOOROODUC ESTATE

Moorooduc Estate, 501 Derril Rd., Moorooduc, Vic 3930, Australia

TO VISIT BEFORE YOU DIE BECAUSE

A Master of Wine oversees this family vineyard.

For many Australians, the Mornington Peninsula, surrounded on three sides by water, is a top beach getaway. But it's also home to standout Pinot Noir, Chardonnay, and Pinot Gris. Leading the charge is Master of Wine Kate McIntyre of Moorooduc Estate. Her title is carried by only 419 people in the world as of 2022. The vineyard was originally planted by her parents in 1983. Sandy topsoils cover deeper layers of clay soils, and as vines get older roots run deep enough to pull nutrients from the earth, eliminating the need for irrigation. Today, Kate shares her knowledge with visitors at the winery's tasting room.

mooroоducestate.com.au +61 (0)3 59718506

144 OMENSETTER VINEYARD – BRASH HIGGINS WINE CO.

242 California Rd., McLaren Vale, Australia

TO VISIT BEFORE YOU DIE BECAUSE

A love story turns into one of Australia's most cutting-edge wineries.

When former American sommelier Brad Hickey came to Australia for harvest, he didn't expect to find love. But he met Nicole Thorpe and they embarked on a personal and professional journey that lead to the creation of Brash Higgins winery. The site of their passion is the Omensetter Vineyard, which Nicole originally planted in 1997. Seeing the potential for the mineral-rich clay and limestone soils beyond the flailing Cabernet and Shiraz, the duo dug into research. The Italian grape Nero d'Avola stood out as a contender and they planted the first of the variety in McLaren Vale. The couple also takes credit for the first and only commercial planting in Australia of Crystal, a little-known Greek/Chinese variety. Book an appointment for this one-of-a-kind tasting.

brashhiggins.com +61885564237

145 PIPERS RIVER ESTATE VINEYARD – JANSZ TASMANIA

1216B Pipers Brook Rd., Pipers Brook TAS 7254, Australia

TO VISIT BEFORE YOU DIE BECAUSE

Its famous sparkling wines come from vines planted by Champagne Louis Roederer.

For the island of Tasmania, the southernmost winemaking region of Australia, strong winds figure prominently in the winemaking history. These blasts of air carried Dutch explorer Abel Janszoon Tasman to the island, and later convinced Champagne Louis Roederer of the island's climactic similarities to the famous French region, owing to the winds' cooling effects. In 1997, the Hill-Smith family purchased the Champagne-inspired vineyard and created Jansz Tasmania, which is now considered one of Australia's top sparkling wines. Get blown away by what's in your glass as you taste wines at a picnic in the vineyard.

jansz.com.au

+61 (03) 6382 7066

146 TARANGA VINEYARD – OLIVER TARANGA'S VINEYARDS

246 Seaview Rd., McLaren Vale 5171, South Australia

TO VISIT BEFORE YOU DIE BECAUSE

This 6th-generation-farmed vineyard grows a cornucopia of varieties.

It remains a mystery why the Scottish-born Oliver family decided to move to Australia in the 1800s, and why they decided to grow grapes without any prior experience. However, the risk paid off. Today, the 100-hectare (247-acre) Taranga vineyard is so well regarded that not only does winemaker Corinna Wright produce award-winning estate wines, she also sells fruit to many of Australia's other top wineries. But it's not necessarily just Shiraz that takes center stage; Italian and Spanish varieties such as Vermentino, Fiano, Tempranillo, Mencia, Mataro, and Grenache also flourish. Taste this wide-ranging portfolio at the recently renovated tasting room or take a golf cart tour through the immense vineyard.

oliverstaranga.com +61883238498

147 LOVEBLOCK FARM – LOVEBLOCK VINTNERS

155 Ugbrooke Rd., RD4, Seddon, New Zealand

TO VISIT
BEFORE YOU DIE
BECAUSE

It's the next phase for one of the most recognizable names in wine.

There's no name more synonymous with New Zealand wine than Kim Crawford's. But although he and his wife Erica sold their famous eponymous label, they were far from ready to retire from winemaking. Loveblock is their highly personal follow-up vineyard and winery that puts organic and biodynamic farming front and center. Situated high on a hill in Awatere Valley, the vineyard was partitioned by the couple into two blocks: Hillside, which serves as an experimental playground for varieties, as well as a key place for aromatic white grapes; and Woolshed (and Triangle) plots on the valley floor. One sip of their aromatic wines and you'll also fall in love.

loveblockwine.com +6493651471

148 GIANTS WINERY AND VINEYARD – CRAGGY RANGE

253 Waimarama Rd., Havelock North 4294, New Zealand

TO VISIT BEFORE YOU DIE BECAUSE

Giant ambitions led to one of New Zealand's most acclaimed wineries.

The Peabody family could have settled anywhere in the world when they decided to start a winery, but the stony and sandy alluvial soils of Gimblett Gravels caught their attention. Their ambitions were indeed giant; they wanted to create a new vineyard, planted to their specifications in order to create legacy wines. Along with the vineyard came Giants winery, the heart of the estate. It has three separate cellars, one solely dedicated to their top wine, Sophia. The winery's hospitality is equally outsized. Dine at the on-site restaurant, admire the sculptures on the property, visit the organic vegetable garden, or take a tour of the cellars.

craggyrange.com +64272794364

149 DRY RIVER ESTATE VINEYARD – DRY RIVER WINES

192 Regent St., Martinborough, New Zealand

TO VISIT
BEFORE YOU DIE
BECAUSE

Here you will delight in the exceptional wines from one of Martinborough's first vineyards.

In its first four decades of existence, the small township of Martinborough flourished with boutique producers, and Dry River was one of the pioneering estates. Pinot Gris was the first to be planted, and these old vines, still on original rootstock, continue to produce concentrated, elegant fruit. Gewürztraminer, Pinot Noir, and Chardonnay also thrive in the vineyard. Adding to the wine's character is a band of clay running through the soil, which is uncommon in Martinborough. It retains water and provides nourishment to the vines. The tasting experience is as intimate as the winery itself. Dry River offers personalized guided tours and tastings with winemakers, by appointment only.

dryriver.co.nz +64 6306 9388

150 TAYLORS PASS VINEYARD – VILLA MARIA WINES

118 Montgomerie Rd., Mangere, Auckland, New Zealand

TO VISIT BEFORE YOU DIE BECAUSE

Here you can climb your way through different soils.

Head to the Taylors Pass vineyard and the ground changes literally beneath your feet. From stony gravel close to the river to deep silt over clay-papa (soft, blue mudstone) at the highest elevation, every terraced level of this renowned site reveals a different type of soil. When it comes to the Pinot Noir and Sauvignon Blanc, the grapes give off different characteristics based on where they grow, allowing winemakers to blend parcels for the ultimate expressions. Get a leg up on your wine knowledge during a vineyard tour, then enjoy a wine tasting or a wine-and-food pairing at the winery's Tasting Room and Cellar Door.

villamariawines.com +64 9 255 1777

Index 150 Vineyards

© Photos

p.12 Darell Lourens / p.13-15 Hamilton Russell Vineyards / p.16-17 Sean Gibson Photograhpy / p.18 Kanonkop wine estate / p.19-21 Klein Constantia / p.22-25 Paul Cluver Family wines & Gibson Photo/ p.26 Pablo Monton / p.27 Bodega Colomé / p.28 Susana Balbo / p.29-31 Piedra Infinita – Zuccardi / p. 32-33 Familia Torres / p.34 Nataschia Wielink Photography / p. 35-36 Clos Apalta / p.38-41 Seña / p. 42 Constanza Miranda / p.43-45 Don Melchor / p.46 Santo Tomas / p.47 @bodegagarzon / p.48 Jon Golden-Barboursville Vineyards / p.49 Jason Tesauro / p.50 Randy Caparoso Photograhpy / p.51 Matt Nelsen / p.52-53 Twain-Peterson MW / p.54 805 Aerial / p.55 Kelsey Chance, kelsey@goodchancecreative.com, 360-224-0625, @goodchance-creative, www.kelseychance.com / p.56-57 Burges Smith, Eisele Vineyard / p.58 L'Ecole N° 41 / p.59 Robert B Richards / p.60 Kim Fetrow / p.61 Jordan Winery Vineyard / p.62-63 Hermann J. Wiemer Vineyard / p.64-65 Stu Gallagher; Dr Konstantin Frank / p.66 Stu Gallagher / p.67 Dr. Konstantin Frank Winery / p.68 Kronos Vineyard, Corison Winery / p.69 Pedernales Cellars / p.70-71 Oak&Barrel Photography / p.72-73 Monte Rosso, Louis M. Martini / p.74 Pisoni / p.75-77 Pym-Rae / p.78-79 Rooftop Reds / p.80-83 Emma K. Moris / p.84 Kim Carroll / p.85 R.J Studio / p.86 Erin Feinblatt / p.87-89 Tablas Creek Vineyard / p.90 Robert Mondavi Winery / p.91-93 Jerry Dodrill Photography / p.94 SW Wine Center / p.95 Tom Schumacher / p.96 Wölffer Estate Vineyard / p.97 Maxence Dulou / p.98-99 Tsinandali Estate / p.100 & p.102-103 Muhkrani / p. 101 Sula Vineyards / p.104 Vitkin Winery / p.105 Yaara Waxelbaum-Stahl / p.106 Lumiere Winery / p.107 Junya Igarashi / p. 108-111 Chateau Musar / p.112-113 Domaine des Tourelles / p.114 Granmonte Co., Ltd. / p.115-117 Yacoubian-Hobbs / p.118 Domaine Wachau Archive / p.119 Gut Oggau / p.120 @GerdKressl / p.121 Philipp Horak for Heid Schröck & Söhne / p.122-123 Wijndomein Aldeneyck / p.124 Bibich Winery / p.125 Njord Vineyard / p.126 Gusbourne / p.127-129 MDRN LOVE / p.130 Philippe Marniquet / p.131 Château Angélus / p.132 Domaines Ott / p.133 Brice Braastad / p.134-135 Alain Benoit / p.136 Pontet-Canet / p.137 Soufiane Zaidi / p.138 JL Bernuy / p.139-141 Domaine Huet / p.142 Alexandre Moyse / p.143 Leif Carlsson / p.144-145 Olivier Wymann / p.146 The Clos Saint Urbain / p.147 Réserve D'o / p.148 Domaine Tempier / p.149 Brigitte Mazéas / p.150 E.GUIGAL ESTATE / p.151 Serge Chapuis / p.152 Tessa Laroche / p.153 Atelier Nicolas Demoulins / p.154 M. Chapoutier / p.155 Champagne Lanson / p.156 Château du moulin-à-vent / p.157-159 Maison Ruinart / p.160 Atelier Nicolas Demoulins / p.161 Kevin Day / p.162 Schoelhammer; Famille Hugel / p.163-165 Brauneberger Juffer; Weingut Fritz Haag / p.166 Verband deutscher Prädikatsweingüter for Weingut von Winning / p.167 @HerbertLehmann for Weingüter Geheimrat J. Wegeler / p.168 Dr. Loosen Estate / p.169 Alpha Estate / p.170-171 Olga Deikou / p.172 Gaia Wines / p.173-175 Bodegas Vega Sicilia for Tokaj-Oremus / p.176 Mézes Maly – Royal Tokaji / p.177 Thilo Weimar / p.178-179 Gianni Bodini / p.180 Colonia; Felsina / p.181 Ornellaia / p.182 ©Abbazia di Novacella; Wolfgang Gafriller / p.183 ©Abbazia di Novacella; Albert Ceolan (top); Werner Waldboth (bottom) / p.184 Fabio Gambina / p.185-187 Faedo; Pojer e Sandri Azienda Agricola / p.188 Elke M. Talbot / p.189 Ivano Cerullo / p.190-191 Il Greppo; Bondi-Santi / p.192 Pio Cesare / p.193 Francesco Rocco / p.194-195 Alessandro di Gresy / p.196-199 Walter Prina / p. 200 Sori Tildin; Gaja / p.201 Kevin Day / p.202 Elena Walch / p.204 Filipa Pato and William Wouters / p. 205 Luis Ferraz / p.206 ©Esporão / p.207 Quinta da Lixa / Monverde Wine Experience Hotel / p.208 Symington Family Estates / p.209 Soalheiro / p.210-211 Quinta de Vargellas; Taylor Fladgate / p.212 Marijan Močivnik/ Studio Ajd / p.213 Ameztoi / p.214-215 CVNE; Contino Estate / p.216 Siboney Tabares King / p.217 Ferran Marti for Clos Mogador / p.218-219 Clos Mogador / p.221 Mas de la Rosa; Vall Llach / p.222 Jesus Rocandio / p.223 Adrian Ruiz de Hierro / p.224 Bodegas Vega Sicilia / p.225 Gonzales Byass Jerez archive / p.226-227 Arilds Vingard / p.228 Sedrik Nemeth / p.230 Vinkara Winery / p.231-233 Giant Steps / p.234 Andre Castellucci / p.235-237 Dragan Radocaj Photography / p.238 Vasse Felix Wines / p.239 Mcintryre Vineyard; Moorooduc Estate / p.240 Brad Hickey / p.241-243 Jansz Tasmania / p.244-245 Taranga Vineyard; Oliver Taranga's Vineyards / p.246 Jim Tannock Photograhpy Ltd. / p.247-249 Richard Brimer / p.250 Robert Brianza / p.251-253 Mike MacKinven

In the same series

Colophon